Business Plan for Sustainable Agri-Business

Business Plan for Sustainable Agri-Business

Authors:

Jyoti Kachroo

Naveed Hamid

Vansh Gupta

SSPH

SATISH SERIAL PUBLISHING HOUSE

403, Express Tower, Commercial Complex
Azadpur, Delhi-110033 (India)
Phone : 011-27672852, Fax : 91-11-27672046
e-mail : info@satishserial.com, hkjain1975@yahoo.com
Website : www.satishserial.com

Published by :

SATISH SERIAL PUBLISHING HOUSE
403, Express Tower, Commercial Complex, Azadpur, Delhi-110033 (INDIA)
Phone : 011-27672852 Fax : 91-11-27672046
E-mail : info@satishserial.com, hkjain1975@yahoo.com

ISBN 978-93-88020-75-6

Composed, Designed & Printed in India

About the Authors

Dr Jyoti Kachroo has remained gold medalist (Two) and meritorious throughout her academic career, graduated and post graduate (Economics) from University of Jammu, Jammu and Kashmir. She was initially appointed as lecturer (Economics) in Govt. Degree Collage Jammu and later joined the division of agricultural Economics and Statistics, SKUAST Jammu as Associate Professor. Presently she is Professor and Head, Division of Agricultural Economics & Agri-Business Management. Her field of expertise is women empowerment, agricultural production economics, marketing and econometric model based research. She has published about more than 90 articles which include chapters, popular articles, review papers and about 65 research publications in referred journals. She has been honored with D.T.Doshi Foundation Award for best presentation. She has more than 26 years experience of teaching and research and she has guided 20 research scholars.

Mr. Naveed Hamid possesses a meritorious academic record and awarded by university merit certificate in recognition of his having achieved the prescribed scholastic standards of high order OGPA in Masters (Agri-Business Management) programme. He has more than 4 years of teaching and research experience in the field of Agricultural Business. He has more than 60 research publications in referred, NAAS rated and UGC approved journals, 22 abstracts, 21 books chapters, several articles, 06 books and so many International/National conferences to his credit. He is recipient of more than 12 International/National recognition award from repute organization for his contribution to the Agri-Business field. Among those awards he is bestowed with Krishi Bhushan and IRDO Young Fellow award. His field of expertise is Agricultural Business Management, Supply chain/Value chain and Marketing. Mr. Naveed Hamid is presently working in Ministry of Agriculture Co-operation and Farmers Welfare. He is also Ambassador, ACPCL (M.P) for promotion of FPO and Agri-business related activities in Jammu and Kashmir State.

Mr. Vansh Gupta possesses a meritorious academic record in Masters (Agri-Business Management) programme. He has obtained his degree in MBA (Agri-Business Management) from the Sher-e-Kashmir University of Agricultural Sciences and Technology of Jammu. He has one research publications in referred and UGC journals. His field of expertise is Agricultural Business Management, Food Industry Marketing. He is presently working in National Seed Corporation, India.

Preface

A business plan provides a complete description of a business, explaining its products or services, marketing and sales strategies, the management team, operations and predictions of financial forecasts. The business plan, therefore, enables a business idea to be taken from the initial conception stage towards a fully reasoned and realistic plan of action. The plan also serves as a working document and essential management tool as it sets out how the business will proceed and the strategies it will employ. It enables possible obstacles to be highlighted and avoided, targets to be focussed upon and achieved, and effective structures to be put in place for finances and business strategies. A business plan is essentially a future oriented description of how one intends to trade as a commercially viable operation.

So, this book focusses specially on knowing the perception of entrepreneurs towards the importance of business plan, to analyze the challenges while running business plan and to analyze the effect of essential features of business plan on the success or failure of a business. The plan should serve as a mechanism to connect the production, marketing and financial aspects of the operation, while at the same time considering the business transition needs, etc. In a very general sense, the plan is a statement of how one intends to react to the constant changes in the business environment in which one operates, and how to achieve the goals and objectives that have set. Most importantly, the process can help to identify the competitive advantage (or advantages) of the individual operation, and help to devise specific strategies and tactics to capitalize on those advantages.

The First chapter of the book focusses on significance of business plan in present era. The Next chapter of this book provides a brief note on Agro Based Industry Scenario and Performance Statistics. The next chapter discusses Business plan for sustainable agri-business. The next chapter reveals about Conclusion of the whole primary study.

We felt our heartfelt gratitude to Dr.Daleep Kachroo (Chief Scientist and Head, FRS), SKUAST Jammu, whose genuine guidance and conceptual technical ideas during the course of planning has timely helped in finalizing this book.

Authors

Contents

Chapter 1

Significance of Business Plan in Present Era

A business plan provides a complete description of a business, explaining its products or services, marketing and sales strategies, the management team, operations and predictions of financial forecasts. In order to demonstrate the viability of the business idea, identifying that the product or service has a definite market, outline the advantages in terms of competition, and appraising the costs and profit margins are the pre-requisites. The business plan, therefore, enables a business idea to be taken from the initial conception stage towards a fully reasoned and realistic plan of action.

The plan also serves as a working document and essential management tool as it sets out how the business will proceed and the strategies it will employ. It enables possible obstacles to be highlighted and avoided, targets to be focussed upon and achieved, and effective structures to be put in place for finances and business strategies. Businesses should use the plan to guide decision making and it should be reviewed, modified and developed as the business progresses and evolves.

Businesses that implement their business plan and keep it up-to-date are able to monitor their growth and are in a better position to seek out external funding to assist with future developments. The plan must develop the procedures and strategies needed to turn a business opportunity into a reality. Furthermore, it must become a reference point to measure the firm's performance during its implementation, by incorporating financial management, social and legal projections.

The guide to the project management body of knowledge defines a project as 'a temporary endeavour undertaken to create a unique product or service'. Taking into account these definitions, there is no doubt that a business plan can be conceived as a project, the first and most important project a start-up must face, to turn a business opportunity into a reality .This shows that projects are non-routine processes that involve the novel coordination of inter-related activities and resources to achieve beneficial change'. Such attributes clearly emphasize the complex, emergent and negotiated nature of project activity as achieved through the collective interactions of project participants and other interested stakeholders.

To understand the concept of a business plan and the business planning process in general, it is important to recognize that the plan is not a document. The plan may be presented through a planning document, but the business plan itself should be a set of management decisions about what a company will be doing to ensure success in the future. In its most basic form, business planning is the process whereby a company makes the following decisions:

- What products or services it will deliver?
- Where will it compete?
- Why the customer should do business with the company?
- How will it organize and re-engineer itself to accomplish its goals?

Business planning is used, to some degree, by all organizations—large and small, for-profit and not-for-profit—regardless of the nature of the business. The scope and complexity of the business planning process and the resulting plan should mirror the size and complexity of the company and its business. For smaller companies that operate in a fairly stable business environment, the planning process may be an informal review of certain key aspects of the company's performance by a handful of top managers. As a company becomes larger, or as the business environment becomes less stable and less predictable, the planning process becomes more formal, comprehensive, and complex. More personnel are involved in reviewing and planning various aspects of the company's performance, and the time horizon addressed by planning decisions may extend further into the future.

1.1 Importance of a business plan

While business plans are synonymous with start-ups, business planning is important for all businesses regardless of their age or maturity. A business plan is essentially a future oriented-description of how one intends to trade as a commercially viable operation. Given that the plan is essentially predictive in nature, it will need to contain a number of assumptions which the reader will need to accept as being plausible in the context of the opportunity being described. While trading, it is a little easier to forecast data as some historic facts upon which one can estimate with greater certainty. However, everybody faces uncertainty in their daily lives which is precisely the reason one needs to embrace business planning.

The process of business planning is, in and of itself, a worthwhile pursuit as it forces the managers to remove themselves from the day-to-day tactical and responsive mode in which they operate. In particular, they must take into account the resources at the company's disposal and plan to maximise the return on capital, with the resources at their disposal.

The following is a list of some examples as to why business planning is so important:

Ensure staffs are focussed on key activities

Manage cash flow

Manage performance

Manage exposure to external conditions

Manage vulnerabilities

1.2 Ensure staff is focussed on key activities

It is important that all employees are aligned and focussed on achieving the same goals. Rather than spending time reacting to emails, phone calls and the like, a company needs to proactively focus on goals and the key tasks which contribute to the key aims of the company. By committing the thoughts to paper, one can understand the business better and also chart specific courses of action that need to be taken to improve the business. A plan can also detail alternative future scenarios and set specific objectives and goals along with the resources required to achieve these goals.

Manage cashflow

Alongside the risk of reduced sales numbers, credit management tends to be the next biggest issue during tough times; debtors take longer to pay, or worse, some debtors go out of business. If the company has a high percentage of sales on credit, it will need to re-consider the payment terms, ensure the invoice immediately and maintain contact with those who owe the money. There is no need to panic; it is simply opportune to revisit all elements of the credit control system, from invoicing, to bank overdraft facilities, to the debt collection process.

Manage performance

A business plan can also be used as a management tool to assess *'actual results' against 'planned results'*. The plan can assign milestones to specific individuals and ultimately help management to monitor progress. By committing the plan to paper, all other options are effectively marginalized and the company is aligned to focus on key activities, with individual responsibilities clearly outlined.

Manage exposure to external conditions

Management surveys the factors that exist independently of the enterprise but which it must consider for profitable advantage. Management also evaluates the relationships among departments in order to coordinate their activities. Some general areas of the external environment considered by management are:

- Demographic changes: sex, age, absolute numbers, location, movement.
- Economic conditions: employment level, regional performance, sex, age, wage levels, spending patterns, consumer debt.
- Government fiscal policy and regulations: level of spending and entitlements, war and peace, tax policies, environmental regulations.
- Labour supply: age, sex, education, cultural factors, work ethics, training.
- Competition: market penetration and position, market share, commodities or niche product.
- Vendors: financial soundness, quality and quantity of product, research and development capabilities, alternatives, foreign, domestic, just-in-time capabilities.

Manage vulnerabilities

The recent economic problems have been particularly painful in certain industries. All business is cyclical. Companies that had been planning their cashflow or sales forecasts were better placed to spot the market decline and would have been best placed to diversify or take more urgent remedial action than those who lack a plan and are essentially rudderless.

1.3 Functions of business plan

Details of a business

Initial costs

Securing financing

Marketing plan

Details of a business

Business plan explains what a business is about by describing the products or services in detail and what the ultimate goals of business are. For example, a plan may stipulate what the revenue goals are for each of first three years of operation. The plan should also indicate why to believe that there is a need for the business and who the main competitors will be.

Initial costs

If the business requires any start-up capital or costs, the business plan should list all required items and the anticipated cost for each. If a person seeking financing, then it will be necessary to determine how much money is needed to borrow. Possible costs can include a place of business, equipment, the hiring of employees and marketing expenses.

Securing financing

A business plan is essential if one intends to seek financing to start a business. A financial institution, such as a bank will want to see the plan to determine the viability of the business idea. If anyone seeking investors as a means of raising capital, they'll also most likely want to view the plan to determine the level of investment risk.

Marketing plan

A business plan should describe how one intends to market the products or services as well as identify the target market. For example, if a firm selling a product that appeals to young professionals between the ages of 25 and 34, the plan should include the marketing strategy the firm intend to use to reach them, such as developing advertising for media outlets that appeal to that particular demographic.

1.4 Need of a business plan

Grow existing business

Back up a business loan application

Seek investment for a business, whether it's a start-up or not.

Create a new business

Valuation of the business for formal transactions related to inheritance, estate planning and tax issues

Sell the business

Deal with professionals

Develop new business alliances

Share and explain business objectives with the management team, employees and new hires.

Decide whether one needs new assets, how many, and whether to buy or lease them.

Hire new people

Decide whether or not to rent new space.

Share strategy, priorities and specific action points with the spouse, partner or significant other.

Set specific objectives for managers.

1.5 Primary parts to a business plan

- **The first is the** *business concept,* where one discusses the industry, business structure, the particular product or service, and how one plans to make the business successful.
- **The second is the** *marketplace section,* in which one describes and analyzes potential customers: who and where they are, what makes them buy and so on. Here, one also describes the competition and how to beat it.

- **Finally, the** *financial section* contains the income and cashflow statement, balance sheet and other financial ratios, such as break-even analyses. This part may require help from an accountant and a good spreadsheet software program.

Breaking these three major sections down even further, a business plan consists of seven key components:

- Executive summary
- Business description
- Market strategies
- Competitive analysis
- Design and development plan
- Operations and management plan
- Financial factors

Types of business plans

Business plans can be divided roughly into four separate types. There are very short plans, or miniplans. There are working plans, presentation plans and even electronic plans. They require very different amounts of labour and not always with proportionately different results.

Mini plan

A mini plan may consists of 1 to 10 pages and should include at least cursory attention to such key matters as business concept, financing needs, marketing plan and financial statements, especially cash flow, income projection and balance sheet. It's a great way to quickly test a business concept or measure the interest of a potential partner or minor investor. It can also serve as a valuable prelude to a full-length plan later on.

Working plan

A working plan is a tool to be used to operate the business. It has to be long on detail but may be short on presentation. As with a miniplan, one can probably afford a somewhat higher degree of candor and informality, when preparing a working plan.

Presentation plan

If taking a working plan, with its low stress on cosmetics and impression, and twist the knob to boost the amount of attention paid to its looks, it will wind up with a presentation plan. This plan is suitable for showing to bankers, investors and others outside the company.

Electronic plan

The majority of business plans are composed on a computer of some kind, then printed out and presented in hard copy. But more and more business information that once was transferred between parties only on paper is now sent electronically. So, one may find it appropriate to have an electronic version of the plan available. An electronic plan can be handy for presentations to a group using a computer-driven overhead projector, for example, or for satisfying the demands of a discriminating investor who wants to be able to delve deeply into the underpinnings of complex spreadsheets.

Objectives and uses of business plan

In general, a business plan is necessary to provide a clear, precise, and meaningful sense of direction for a company for a specific period of time—usually 1 year. The preparation of an annual plan forces management to think through its intended actions and shows how they will affect various aspects of the business. Beyond these overriding purposes, a business plan can have several different objectives that, in turn, relate very closely to the plan's intended uses—both internal and external.

Internal uses of a business plan

- It can improve performance by identifying the strengths and weaknesses of the company's operation and any potential or emerging problem areas. Improved performance also will result from developing strategies that simultaneously build on the company's strengths and correct its weaknesses.
- A business plan can communicate to management and staff clear expectations regarding the company's performance and priorities.

- For companies with multiple divisions, units, or points of management responsibility, a business plan can effectively coordinate and ensure consistency in the plans and operations of the various units or divisions.
- A business plan provides a solid basis for measuring performance of the overall company and of individual units and managers. It establishes a standard for deciding if actual performance is good, bad, or indifferent. A business plan also establishes a framework for reviewing key decisions in the ongoing management of the company. As new developments and opportunities arise, a business plan provides a rational structure for evaluating their impact on the operations and performance of the overall company and its individual units.
- The most important, a business plan and the process of developing it can be used to educate and motivate the key staff—such as managers and supervisors—of the company. Through participation in analyzing past performance, evaluating the impact of trends and developments, and developing action plans for the future, managers and other key staff can learn more about the total operation of a company and the relationship of their specific areas of responsibility to the achievement of the company's overall goals.

External uses of a business plan

- It can be used to educate outside parties regarding the objectives, structure, and performance of the company. This use becomes important if the support, either financial or non-financial, of outside parties is important to the company's success. A clear example is the support of current or potential stockholders or investors.
- A business plan can be used to secure funding from outside investors, either individual or institutional. If this use is intended, it is important to identify in advance the issues and types of information that are of primary concern and interest to the investors.

A business plan also can communicate planned actions, deflecting competitive or regulatory moves that may be under consideration by outside parties. By formalizing its intentions in a business plan, a company commits itself to action. By communicating this commitment to selected, significant outside parties, a company may pre-empt actions that could have a negative impact on its own plans and actions.

Given different objectives and uses of the business plan, management must decide whether it is desirable or necessary to distribute its plan. If the plan is to be distributed, the specific structure and content will be affected by the intended uses and target audience.

Qualities of a good business plan

- A business plan should be detailed. In listing the products and services for example, one should not really stop by just enumerating them. One also should write down the descriptions and scope of his products and services, touch base on production and identify means on how to market the targeted niche.
- It should include a market research that identifies the competitors, share of the market and the range of the products the business unit produce. By learning how to conduct their operations, one may learn tricks of the trade in the business to enter and also get to have a basis on what one can do to excel.
- It needs to have a list of equipment, technology, raw materials, financial and other resources that may need when starting and running the business venture. Having all these listed will give an idea on how much capital one need before starting and how much money should be made in a day to make the business survive.
- It also needs to be written in formal format and style. One has to remember that a business plan is something that may have to present to business partners, financial firms and banks.
- Finally, a business plan should be error free. This is important because a business plan defines one as what kind of a business person. If it turned out sloppy, then that does not speak too highly.

Four critical management processes

- Translating the vision/mission (clarifying, strategy setting and gaining consensus).

- Communicating and linking (communicating/educating, setting business goals, linking with the objectives of the various parts/departments of the organisation).
- Business Planning (by all parts of the organisation, setting targets, aligning strategic initiatives, allocating resources, establishing 'milestones').
- Feedback and learning (articulating the shared vision/mission, supplying strategic feedback, facilitating strategy review and learning).

Major business strategies

The major business strategies which are the outcome of a business plan are:

Market strategy

Production or service strategy

Research and development (R&D) strategy

Organization and management strategy

Financial strategy

Market strategy

A market strategy addresses the following critical issues:

- What products or services will the company sell?
- Who are, or will be, the target customers of the company?
- What will be the company's policies regarding product or service pricing, advertising and promotion, sales, and distribution?

By developing a market strategy, an organization is able to prepare estimates of revenues and selling costs—but not production costs—for its products and services.

Several very important analyses underlie the development of a market strategy, including the following:

- Preference, consumption, and purchasing behaviour—that is, timing and volume—of customer segments
- Usefulness of products or services to customers

- Satisfaction of customers with products or services
- Degree of competition within the industry for the same customer segments

These analyses are important in developing realistic estimates of what can be sold to whom, when, in what quantities, and at what price, given the level of promotion, sales, and distribution effort.

Production or service strategy

A production or service strategy addresses several critical issues: the process and technology that will be used to produce and deliver the company's products or services; the requirements for materials, equipment, and facilities; and the production schedule required to support sales goals. It enables the company to determine estimates of operating income profitability and to determine the level of sales and revenue that the organization must attain before it can become profitable—that is, the breakeven point.

Similar to the market strategy, several important analyses underlie the development of a production or service strategy:

- Identification and evaluation of both the fixed and variable production costs for each product or service.
- Evaluation of required lead times for acquiring materials and the time required for the actual production process.
- Assessment of supplier relationships regarding availability, quality, and cost of materials and services.
- Evaluation of production efficiency and quality control.
- Assessment of production cost efficiency and its impact on overall production costs.

These analyses are important to ensure that sales and revenue objectives are appropriately meshed with the production or service capability of the company.

Research and development strategy

The R&D strategy addresses two key questions: What critical areas of the company's business operation must be developed, maintained, or expanded to sustain a competitive edge? Given trends in the marketplace and the operating environment of the company, what investments are necessary to maintain or expand the competitive edge?

The need for a formal R&D strategy is determined, to a large degree, by the structure of the company's operating environment. If the company operates in a highly competitive environment, characterized by rapid and frequent technological change and requiring medium- to long-range planning horizons, a formal R&D strategy is an operating necessity. If the company operates in a more stable environment or in one where new product introduction is not a primary determinant of market success, a formal R&D strategy is less necessary. As the economy becomes more and more knowledge-and information-based, fewer and fewer companies will have the luxury of not developing some type of R&D strategy.

The need and the scope of an R&D strategy flow from the market analysis. If the market analysis indicates that the purchasing behaviour of target customers is influenced by factors other than price, the need for some level of R&D effort is indicated. The level of effort can range from long-term new product planning and introduction to fine-tuning adjustments of existing products and services.

Organization and management strategy

An organization and management strategy addresses several critical issues:

- Functions that must be performed and who will be responsible for performing them
- Organization of functions to promote overall effectiveness, efficiency, and productivity
- Management of the organization's processes and practices to ensure quality products or services
- Staffing requirements in terms of numbers, skills, qualifications, and promotional opportunities
- Procedures for guiding, motivating, and controlling the overall operation of the company

Through the organization and management strategy, the company begins the process of deciding how to implement its major business strategies. Who is in-charge and who will do the work, particularly regarding critical functions, are important concerns to outside investors.

Several important analyses underlie the development of an organization and management strategy. These analyses are important to ensure that the company effectively addresses the "people" issues that are central to any company's success:

- Evaluation of management attitudes and beliefs regarding company values, the treatment of employees, and key measures of company success
- Availability and cost of required staff capabilities, both within and outside the company—for example, consultants and board members
- Historical strengths and weaknesses in the structure, organization, and management of the company, including communications, inter unit co-operation, and employee morale and productivity.

Financial strategy

A financial strategy addresses three critical issues:

- How business operations will be managed financially to generate an adequate profit or to ensure ongoing economic viability,
- How excess revenues or profits will be used,
- How basic operating funds and funds required for key investments will be generated.

In effect, the development of a financial strategy becomes a reality test for the other strategies in the business plan. A financial strategy answers the overall question of what investments or expenses can be made or incurred.

A financial strategy answers questions, such as the following:

- Does the planned level of operations generate a sufficient level of revenues or profits to support the investment?
- Can the level of planned operations generate a sufficient cash flow to sustain operations? If not, can the shortfall be covered from other sources?
- Do the levels of investment (expenses) and returns (revenues) maintain economic solvency and viability?

As with the other strategies, several analyses underlie the development of a financial strategy. They establish a framework for evaluating the financial viability of the marketing, production or service, R&D, and organization and management strategies. The analyses include the following:

- Historical information regarding the time relationship between the incurrence and payment of costs and the receipt of revenues – that is, the cashflow problem
- Policies of the company and its suppliers and customers regarding payments and collections
- Historical performance of the company on key financial indicators
- Composition of assets and liabilities, including future financial obligations
- Value and current usefulness of assets
- Cost and availability of funds from external sources.

Process of a business plan

Organizing the planning process

Diagnosing the situation

Setting goals

Developing operating plans

Developing a financial plan

Writing the business plan document

Organizing the planning process

Many business planning efforts fall flat because inadequate attention is given at the outset to organizing and energizing the planning activity. Several considerations are involved.

- A company's management must define and clearly communicate its commitment to the planning effort. This commitment should include a statement of the objectives and anticipated uses of the business plan.
- All key parties must be involved in the planning process. Top management should identify who will be involved, including outside parties; determine the appropriate form and timing of

the involvement; provide the opportunity for proper participation; and ensure that participation occurs.

- Responsibilities for developing and implementing the plan should be precisely defined and assigned.
- The steps and process for completing the plan should be timed and made an integral part of the work schedule. An overall time table for developing the plan, including intermediate milestones, should be established. Sufficient time should be incorporated into the time table so that staff can effectively perform both planning and ongoing work responsibilities.

If these considerations are adequately addressed in the organizing phase, the planning process becomes a high priority and a feasible activity for the company and its staff. The result is a high-quality business plan with shared ownership.

Diagnosing the situation

The first analytical step in the planning process is to assess and evaluate what a company is currently doing in relation to its operating environment.

The first part of this analysis is to assess the company's current capabilities— that is, its strengths, weaknesses, and historical performance.

The second part is to assess external marketplace developments— that is, threats and opportunities—and their impact on the future performance of the company.

Information from both parts of the analysis is critical to an effective planning process. For some companies, it is important to monitor external developments continually and evaluate performance in light of responsiveness and adaptability to any developments. These companies typically have the resources as well as the desire to invest in a high level of research and analysis. Other companies prefer to evaluate marketplace developments in light of their specific impact on current performance and capabilities. This approach, which requires significantly less effort and resources, reflects a lower perceived need for timely external review. The specific approach that a company takes depends on the competitive setting, management's preferences, the availability of resources to support ongoing evaluations of the external environment, and the frequency and significance of external developments on the company's performance.

Setting goals

The situational diagnosis produces a set of findings about what is being done and an assessment of the adequacy of existing strategies in terms of current performance and future developments. The results of the situational analysis are used in two separate, but related, ways in the goal-setting step of the planning process. First, the results define specific areas and levels of improvement in the company's operations and management. They also establish performance standards for each key dimension of the company's operations and management. By setting goals, the company is defining the parameters of its overall business strategies for the future.

Within each strategy—marketing, finance, and so forth—goals should be established for each operating dimension in which performance improvement is necessary or appropriate. Goals should address both current and potential strengths and weaknesses, as well as future or emerging opportunities. Across all strategies, goals should be supportive of each other and mutually attainable.

Typically, goals are first established for each strategy area. When viewed collectively, some of these goals may conflict with those of other strategies. Conflict often occurs with marketing and production goals, or with R&D and financial goals. For a business plan to work, conflicting goals must be reconciled and a consistent overall goal framework established.

Because assumptions on which the plan may be based are not perfect predictors of the future, the company should develop alternative goals and strategies based on other feasible assumptions. These alternatives are the bases for contingency planning. Adequate attention to these considerations should yield a good, feasible set of goals and performance standards for the company.

Developing operating plans

Operating plans should be developed for each of the operational strategies that is, marketing, production or service, research and development, and organization and management. The purpose of this step is to translate the strategies into specific business and operating plans for each of the operating components of the company. Each component's objectives and strategies will vary somewhat, but the operating plan for each component should spell out in fairly specific terms the performance objectives for the component and provide a clear, overall indication of how the objectives will be accomplished.

Developing a financial plan

The development of a financial plan flows from the performance objectives and strategies of each of the operating components. These objectives and strategies provide estimates of revenues and expenses. A financial plan must use this information to produce several proforma and supporting schedules that can be used to evaluate and manage the financial performance of the company. Proforma schedules are standard accounting statements prepared by the company's accounting department and are minimum requirements for internal management and external users. The three proforma schedules are as follows:

- Quarterly, and frequently monthly, income statements for the period addressed by the business plan, reflecting the relationship among revenues, expenses, and profits generated
- Quarterly and annual balance sheets, reflecting the relationships among short- and long-term assets and liabilities and owners' equity or level of investment
- Monthly cashflow summaries, reflecting the impact of business operations on the cash position of the company

Supporting schedules provide a more complete description or explanation of the financial figures included in the proforma schedules. Two of the more common supporting schedules are the monthly sales and revenue projections. These projections are obtained from the marketing operating plan and monthly departmental budgets that reflect planned expenses and expenditures for production and overhead operations. Beyond these schedules, a financial plan should include summaries of planned financial performance according to selected indicators of profitability, growth, solvency, and return on investment.

The results of a financial analysis provide an indication of the adequacy and attainability of the operating objectives and strategies from the perspective of available financial resources. Based on the results of the financial analysis, it may be necessary to re-evaluate and revise the operating objectives to support more satisfactory and viable financial performance.

Writing the business plan document

The preparation of a formal business plan document is a requirement, if external uses of the plan are anticipated. A formal document should be prepared, even if the intended uses of the business plan are strictly internal, in order to provide a written record of the management decisions that have been made during the planning process.

Preparing the planning document has three objectives. First, it brings closure to the planning process. Second, it communicates the seriousness of the planning process and a commitment to resultant objectives and strategies. Finally, the planning document provides a tangible framework for evaluating subsequent activities, accomplishments and opportunities.

Preparing the business plan document is an exercise in writing, but not in creative writing. The document should simply and concisely present the analyses and decisions that have been made during the planning process. In general, the longer the planning document is and the more verbiage it contains, the more the document will reflect decisions that have not been made, rather than those that have.

The document should be organized according to the major strategies of the business plan or at least those strategies that are relevant to the company's operation. The document should begin with an executive summary, which briefly highlights each major section. The document also should include a brief assessment of the company's market position and business outlook. The assessment is important to validating the assumptions that will be reflected in the business plan strategies.

Strong reasons for having a good business planning process in a business

Planning gives	Planning helps
Certainty	Shape thoughts
Confidence	Think through scenarios
Route map	Co-ordinate activities
Evidence to others of one's thoughts	Identify SWOT
	Examine risks
	Communicate ideas

❏❏❏

Chapter 2

Agro-Based Industry Scenario and Performance Statistics

The agro-based industries have huge base in india because agricultural activities contribute about 13.7 per cent to our national gross domestic products (GDP) and about 47 per cent of labour force is employed in the agriculture (www.indiainbusiness.nic.in). The business plan plays a very important role in the fast growth and the success of agro-based units in India. The agro-based units are also growing in Jammu region at a very fast rate, the study is to help the small scale agro-based industries to achieve their goals by adopting a well-structured business plan and how business plan help them to manage the company, attract the finances and achieve future targets by providing a complete description of a business, explaining its products or services, marketing and sales strategies, the management team, operations and predictions of financial forecasts.

2.1 Agro-based industry in India

India is one of the world's largest producers of food, and is the largest producer of milk, sugarcane and tea, as well as the second largest producer of rice, wheat, fruits, and vegetables. Nearly 70% of the population depends on agriculture and agro-based industries. (www.wikipedia.com). The agro industry is regarded as an extended arm of agriculture. The development of the agro industry can help stabilize and make agriculture more lucrative and create employment opportunities both at the production and marketing stages. The broad-based development of the agro-products industry has the capacity to improve both the social and physical infrastructure of India. Since it causes diversification and commercialization of agriculture, it can, thus, enhance the incomes of farmers and create food surpluses.

The agro-industry mainly comprises of the post-harvest activities of processing and preserving agricultural products for intermediate or final consumption. It is a well-recognized fact across the world, particularly in the context of industrial development that the importance of agro industries is relative to agriculture increases as economies develop. It should be emphasized that 'food' is not just produce. Food also encompasses a wide variety of processed products. It is in this sense that the agro industry is an important and vital part of the manufacturing sector in developing countries and the means for building industrial capacities. The agro industry is broadly categorized in the following types:

- Village industries owned and run by rural households with very little capital investment and a high level of manual labour; products include pickles, papad, etc.
- Small scale industry characterized by medium investment and semi-automation; products include edible oil, rice mills, etc.
- Large scale industry involving large investment and a high level of automation; products include sugar, jute, cotton mills, etc.

The agro industry helps in processing agricultural products, such as field crops, tree crops, livestock and fisheries and converting them to edible and other usable forms. The private sector is yet to actualize the full potential of the agro industry. The global market is mammoth for sugar, coffee, tea and processed foods, such as sauce, jelly, honey, etc. The market for processed meat, spices and fruits is equally gigantic. Only with mass production coupled with modern technology and intensive marketing can the domestic market as well as the export market be exploited to the fullest extent. It is, therefore, imperative that food manufacturers understand changing consumer preferences, technology with modernization, innovation and incorporation of latest trends and technology in the entire food chain as well as agro-production, the total production capacity of agro products in India and the world is likely to double by the next decade.

India is the second largest producer of food in the world. Whether it is canned food, processed food, food grains, dairy products, frozen food, fish, meat, poultry, the Indian agro industry has a huge potential, the significance and growth of which will never cease.

Performance of Agro-Based Industries

The total number of agro-based industries has increased from 28,584 in 2006-07 to 30,514 in 2011-12 with only 0.38 per cent growth rate. The manufacture of food products and beverages accounts for more than half of the total output of agro-based industries in India. Under this category, the manufacture of grain mill products, starches and starch products and prepared animal feeds is playing a prominent role with a share of 53 per cent. The number of textile manufacturing units has the highest percentage next to food products and beverages, put their number has decreased slightly from 2006-07 to 2011-12, due to the continuous decrease in spinning, weaving and finishing of textiles from 27.25 per cent in 2006-07 to 26.68 per cent in 2011-12. (Annual Survey of Industries, Central Statistical Organization)

In the case of manufacture of food products and beverages, the Compound Growth Rate (CGR) is 5.10 per cent for both beverages, and dairy products and it is 0.48 per cent only for manufacture of other food products category. The CGR is computed for the different categories of agro-based industries, which are significant at 5 per cent level, except manufacture of knitted and crocheted fabrics and articles.

Agro-based industry in Jammu

Jammu and Kashmir mostly has manufacturing industries, small-scale agro industries, etc. There are industries in almost all parts of Jammu and Kashmir but some areas have been marked as primarily and significantly industrial areas. Some of these important areas are:

- Industrial Growth Centre in Samba
- Integrated Infrastructure Development Project in Udhampur
- Industrial Complex in Bari Brahmana

The small scale agro-based industries plays a very important role in making the state favourable for the outside industrial groups so that they come and invest in and generate the existing businesses.

The main types of small-scale agro industries in Jammu are:

- Food and beverage industries
- Sericulture
- Chemical industries
- Bee farming

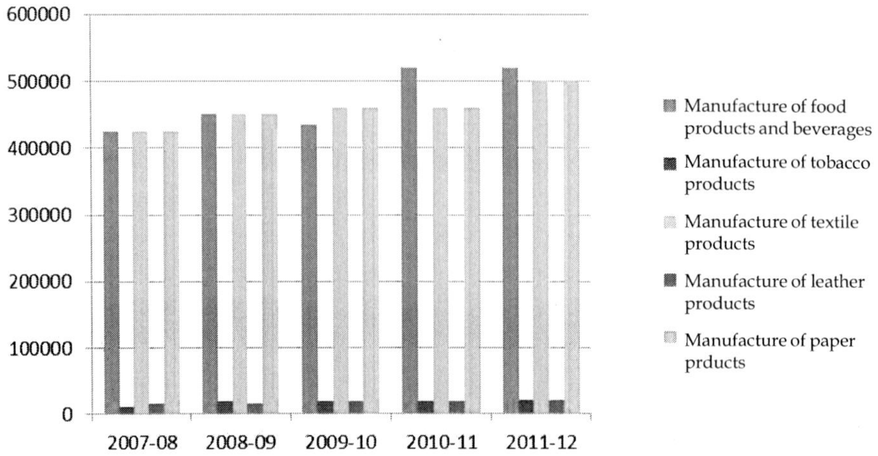

Fig. 2.1: Category-wise agro-based industries in India (number of units)
Source: Annual Survey of Industries, Central Statistical Organization, New Delhi.

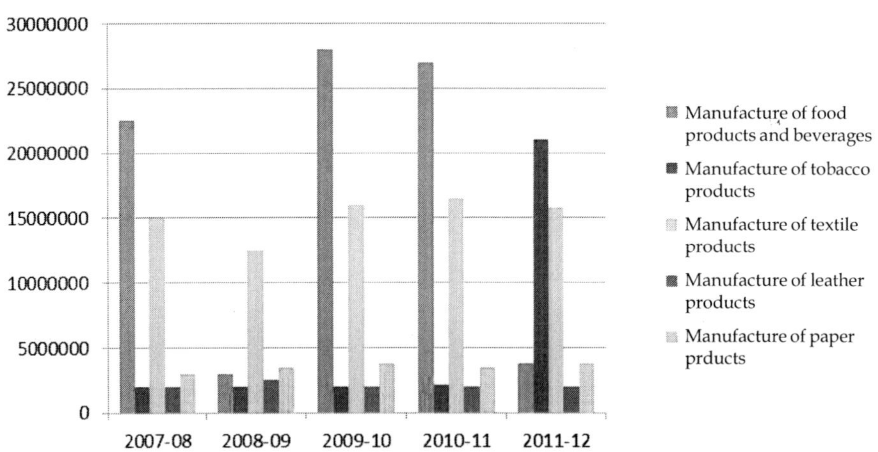

Fig. 2.2: Category-wise working capital of agro-based industries in India (in Lakhs)
Source: Annual Survey of Industries, Central Statistical Organization, New Delhi.

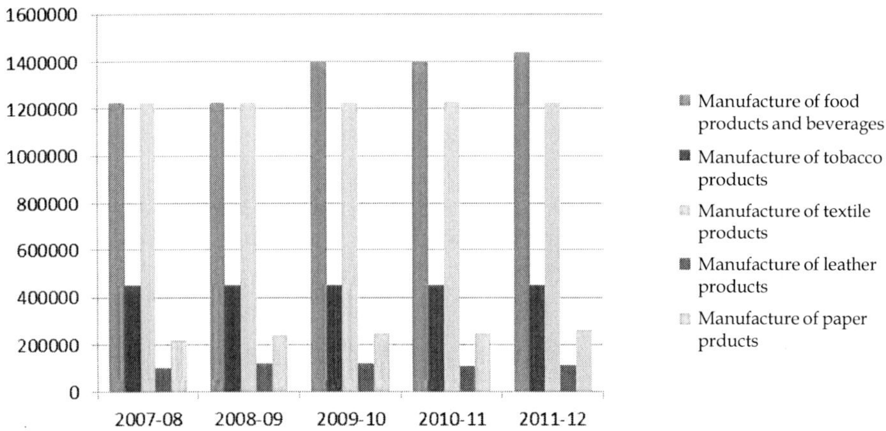

Fig. 2.3: Category-wise net value added of agro-based industries in India (in Lakhs)
Source: Annual Survey of Industries, Central Statistical Organization, New Delhi.

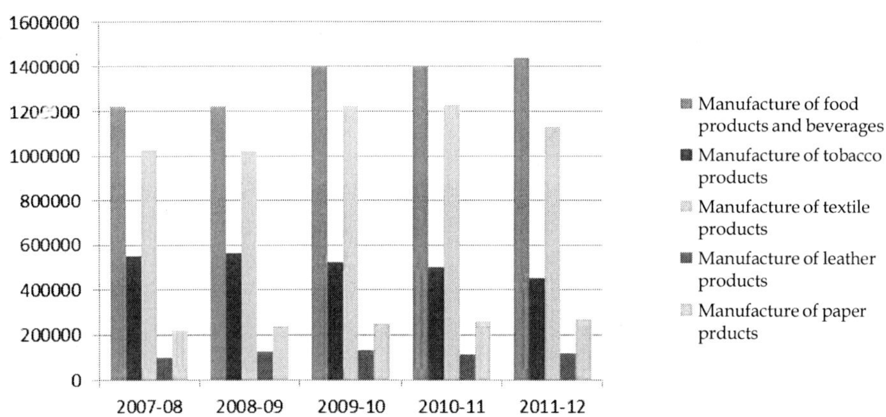

Fig. 2.4: Category-wise net income of agro-based industries in India (in Lakhs)
Source: Annual Survey of Industries, Central Statistical Organization, New Delhi.

Industrial infrastructure

The State has 50 existing industrial estates spread over an area of 34705.37 Kanals of land ending Nov., 2012. Profile of these Industrial Estates is presented in table 2.1.

The existing infrastructure in most industrial estates spread in the state is dilapidated and there is an urgent need for massive upgradation. During the last financial year up-gradation of following Industrial Estates were taken up by the State under Cluster Development Programme with MSME with the following details as per table 2.2.

Growth in small scale industrial units

For speedy growth of Small Scale Industry sector in the state, the Government has brought about simplification in the registration procedure. One District Industries Centre has been set up in each district of the state. The primary objective of the District Industries Centres is to provide all services to the entrepreneurs under one roof. They serve as the nodal agency in the district for the orderly development of the industries. The District Industries Centres have been entrusted with the responsibility of providing all approvals/ clearances needed for setting up an industry. The SSI Units are initially given the provisional registration and after commencement of production they are given formal registration. The District Industries Centres are also arranging required credit facilities. The total number of permanently registered SSI units ending March 2012 stood at 54,716, generating 2,58,030 employment opportunities. Out of these formally registered units, the number of SSI units is 659 during the current year ending Nov., 2012, providing employment to 4707 persons.

Status of SSI in J&K state

As per third all India census report, 2001 the total number of functional units in the State were 15,253 generating employment for 50,643 persons. The annual turnover of these functional SSI units was estimated to be 2062.27 crores with an investment of 937.56 crores. Year on year, growth is encouraging. On an average, more than 4 persons are employed in each MSME's industry.

Table 2.1: Existing industrial estates in the state

Managing agency	Industrial Estates (No)	Area (kanals)	Units setup (No)
1	2	3	4
Directorate of Industries & Commerce	31	2214.37	842
J&K SIDCO	10	25,124.00	1287
J&K SICOP	9	7367.00	1209
Total	50	34,705.37	3338

Source: Jandkplanning.com/economic-survey

Table 2.2: Existing industrial estates taken up by the state with MSME

S.No	Name of Industrial Estate	Vetted by SIDBI	State share	Central share
1	Industrial Estate, Bari-Brahmna, Samba	1107.69	607.29	500.40
2	Industrial Estate, Gangval Jammu	983.92	486.32	497.60
3	Industrial Estate, Khunmoh, Phase 1 & 2	1154.85	681.07	473.78

Source: Jandkplanning.com/economic-survey

Table 2.3: Growth in small scale industrial units of J&K state

Year (As on March)	No. of SSI registered	Employment generated (No)	%age increase in	
			SSI units	Emploment
2001-02	43689	193285	-	-
2002-03	44701	197578	2.31	2.22
2003-04	45672	203428	2.17	2.96
2004-05	46818	209322	2.50	2.89
2005-06	48224	219127	3.00	4.48
2006-07	49426	225963	2.49	3.11
2007-08	50472	230552	2.12	2.03
2008-09	51443	235918	1.92	2.32
2009-10	52631	243411	2.25	3.00
2010-11	53546	249188	1.74	2.37
2011-12	54716	258030	2.19	3.55

Source: Jandkplanning.com/economic-survey

Table 2.4: Status of SSI sector in the state

(₹ in crores)

Year turnover	No of functional units	Emp. Generated (No)	Investment	Annual turnover
As per 3rd all - India Census 2001	15253	50643	937.56	2062.27
2001-02	973	5485	198.54	231.52
2002-03	1012	5053	466.47	496.25
2003-04	971	5090	247.79	321.96
2004-05	1146	5894	381.71	1319.50
2005-06	1406	9805	128.29	180.27

2006-07	1202	6836	101.81	1493.64
2007-08	1046	4859	82.53	Not Available
2008-09	971	5366	145.01	
2009-10	1188	7493	198.00	
2010-11	915	5777	176.76	
2011-12	1170	8842	408.77	
2012-13	659	4707	194.65	

Source: Jandkplanning.com/economic-survey

2.2 Objectives of the study

- To study the perception of entrepreneurs towards the importance of business plan
- To analyze the challenges while running business plan
- To analyze the effect of essential features of business plan on the success or failure of a business.

2.3 Importance of the study

The plan should serve as a mechanism to connect the production, marketing and financial aspects of the operation, while at the same time considering the business transition needs, etc. In a very general sense, the plan is a statement of how one intends to react to the constant changes in the business environment in which one operates, and how to achieve the goals and objectives that have set. Most importantly, the process can help to identify the competitive advantage (or advantages) of the individual operation, and help to devise specific strategies and tactics to capitalize on those advantages. Specifically, the planning process helps to:

Identify goals of the operation

Availability of inventory resources

Assess the business and the environment that it exists in.

Analyze the business performance

Decide on actions for the operation

Implement strategies

Evaluate the plan

2.4 Scope of the study

The report is beneficial both for the students and the entrepreneurs. It has provided complete information about the factors influencing the success of a business plan and has shown the importance of the business plan among small firms in the Jammu and Kashmir state. It has further helped in establishing a platform for further research. Gathering information about the challenges, followed by entrepreneurs while operating & running business activities is a vital issue that is to be analyzed by taking into consideration the importance of various activities that play a significant role in running the enterprise. Further, the study has focussed on understanding concepts of a business plan.

2.5 Limitations of the study

- Some owners were not willing to give appointment due to their busy schedule.
- The personal bias of respondent was also a limitation.
- Respondent's ignorance to certain questions has posed a barrier towards certain responses.
- Why plan when the future is uncertain anyway.

Studies related to the research problem

A widely-held business school mantra is that a new venture should start with a written business plan. Business plan advocates state that for an aspiring entrepreneur a written business plan has many benefits including articulating the business model, strategic, operational, and financial planning, examining critical assumptions, and fund-raising. While there is no argument in the prescriptive literature on the importance of having such a plan in approaching investors, there is much more controversy concerning a business plan's practical value.

Empirical studies into the effectiveness of business plans have produced mixed results. But there is general agreement that two important reasons for writing a business plan are institutional and mimetic pressure. When those two factors are combined, it is expected that would-be entrepreneurs who are seeking external funding and have taken a basic entrepreneurship course are more likely to write a business plan than those who are not seeking external funding and have never taken an entrepreneurship course.

Importance of business plan

Armstrong (1982) noted that business planning, particularly financial planning, can save entrepreneurs' time and money in the start-up process. Business plans also help entrepreneurs figure out who their potential customers are, what kind of customer needs are out in the market, and how the firm will reach customers. Business planning also helps to map out the firms operations and help the entrepreneur how to produce the good or deliver their service.

Bhide (1994) added depth by suggesting that the impact of business plans on new venture performance is unlikely to be a generic positive, negative or negligible effect. Instead, he argues that the efficacy of business plans are likely to be governed by the context within which, business plans are written. Some are written to raise loan finance with the purpose of re-assuring lenders of the low risk and secure positive cashflow position of the business; others are written to help a founding self-funded entrepreneur devise a market entry and growth strategy for a high risk innovative new product in an emerging uncertain market. The effects on performance are unlikely to be the same in such widely varying contexts.

Mintzberg (1994) said the key to understanding business planning is the concept of formalization. He defined strategic planning as "a formalized procedure to produce an articulated result, in the form of an integrated system of decisions".

Wilkinson and Monkhouse (1994) defined business planning as "a method used to position an organization, through prioritizing its use of resources according to identified goals, in an effort to guide its direction and development over a period of time". Talk of a "disciplined effort," a "formalized procedure," and "a method," points toward the idea of a process and, as such, the discussion begins with a review of the business planning process.

Castrogiovanni (1996) environmental uncertainty stimulates planning because more learning is desired under uncertain conditions than under certain ones. The successful outcomes of actions are not assured when the environment is uncertain, and proactive learning through planning can help the founders to face the challenging environmental conditions easier than without planning.

Drucker (2001) suggested that large and small firms act on the basis of a business theory, which is formed from a group of assumptions on what business the firm is in, what its objectives are, how it defines business results, which its customers are, what they value, and for what they are willing to pay. He further concluded the role of strategy is to put the business theory in action and help the firm to achieve desired results despite the environmental unpredictability. Strategy helps the firm in the purposeful search for opportunities.

Stone and Pemberton (2000) defined that strategic planning as formulation of broad and flexible plans that set out the objectives, strategies and policies that drive the organization towards its vision of future.

Delmar and Shane (2003) planning fastened a decision-making process compared to trial-and-error learning where a rapid discovery must take place in order to survive. By gathering and analyzing information, founders can enhance market understanding and test their assumptions without implementing time-consuming processes of first taking action and expending resources. Hence, resource expenditure can also be diminished; by planning first and then acting afterwards enables entrepreneurs to test assumptions on the paper rather than in action. Also, according to the principles of organization theory, planning before taking action improves the quality of most human action.

Carthy *et al.* (2003) studied that business planning is a wide topic and this literature review deals with major debate around whether the business planning process needs to be prescriptive (deliberate) or emergent within small businesses.

Bryson (2004) defined business planning as "a disciplined effort to produce fundamental decisions and actions that shape and guide what an organization is, what it does, and why it does it".

Shane and Delmar (2004) used a random sample of 223 Swedish entrepreneurs to examine their organizing efforts including business plan completion. Their results showed that entrepreneurs were less likely to fail, if they completed a business plan before beginning market activities.

Acs *et al.* (2005) suggested in their contribution that start-ups play a crucial role in job creation, economic growth, innovation and competitiveness. Ruano (2001) and Garcý´a *et al.* (2002), emphasized their importance in job creation, compared to that of large companies. The work of Ruano (2001) further suggested that the rate of job creation is inversely proportional to the size of the firms.

During the formulation of a business plan, Lane *et al.* (2005) believed that the planning process requires a qualitative methodology to gain a more complete understanding of the behaviour and feelings of the stakeholders involved. They also suggested using multi-method data collection including observation, document collection, and informal interviewing.

Liao and Gartner (2006) found the firms that completed a business plan were 2.6 times more likely to launch their business than those that did not complete a plan.

Gartner & Liao (2009) besides benefitting company founders internally, pre-startup planning generates an external purpose as well. A written plan presents the knowledge of the founders to others in order to gain support and resources. Consequently, a written plan has the benefit to obtain financing and other prospective individuals who are interested in to be involved in the venture, such as suppliers, customers and employees.

Franco *et al.* (2010) confirmed that the literature on business plan activities and its importance typically takes into account three groups of explanatory variables involving those relating to the characteristics of the entrepreneur, those relating to the characteristics of the newly founded firm and those external factors embracing the geographical and industrial environment in which entrepreneurial phenomena occur.

Brinckmann *et al.* (2010). conducted a meta-analysis on the business planning–performance relationship and specifically focus on contextual factors moderating the relationship. Results indicate that planning is beneficial, yet contextual factors, such as newness of the firms and the cultural environment of firms significantly impact the relationship. Based on this evidence, it was proposed a concomitant and dynamic approach that combines planning and learning.

Alexander (2013) in the study found the impact of the business plan on the development of small and medium size enterprises in the food processing sector in the Republic of Macedonia and its importance in the modern competitive market environment. Among economists and entrepreneurs a widely spread opinion is that the greatest impact on economic development of a national economy are small and medium size enterprises. The successful development of small and medium size enterprises requires detailed analysis of the market structure, industry, product and competition. The most detailed analysis of the previously mentioned elements is provided by a business plan that should lead to clarification of some aspects unknown to the original understanding of the directions of the enterprise.

Challenges of business plan

Mintzberg (1978) identified that some strategies are not clearly defined; they emerge, rather, as decisions are made and organizations change. These emergent strategies may not be defined in the planning process, because they are unknown at the time, but still may have an important effect on the success of the organization. The challenge for planners is to remain open to strategies that may emerge, and opportunities and threats that may exist, even though they were not identified through the planning process.

Bracker et al. (1988) planning enabled proactive learning, but cannot lead to enactive learning, in order to learn from a business planning process internally, planning processes must be well-developed, properly implemented and controlled. The process of creating the plan is a key component of performance, not the plan itself that only happens when people create knowledge out of their experiences.

Bruderl et al. (1990) suggested the four main reasons that explain the greater risk of failure that one firm faces were first they depend on new roles and tasks that have to be learned at certain costs, sometimes roles have to be developed, and this may be in conflict with constraints on resources or creativity, social interactions in a new organization resemble those between strangers and stable links to clients, supporters or customers are not yet established, when an organization begins its activity.

Mintzberg (1990), forecasting the future with a complete confidence is not possible, and thus managers should not evaluate their organization's strengths and weaknesses until they become evident from the trial and error experience.

Bhide (1994) concluded the development of a business plan is an important element in the success of the entrepreneurial venture. However, most entrepreneurs and small businesses do not base their success on business plans, because they tend, to a large extent, not to formulate them.

Floyd and Woolridge (1996) created a view that strategy as a field of enquiry developed from a practical need to understand reasons for success and failure among organizations. This led to focus on overall performance and on the top management. Thus, strategy is made at the top and executed at the bottom, further reinforcing the fields focus on the top management while implementation was seen as secondary.

Khusro (1999) suggested that "if an organization attempt to create only employment without regard to efficiency, output and surplus, it will soon end up with neither employment nor output or surplus". Accordingly, Khusro further suggested formulation of a strategy that depends on "self-financing surplus generating schemes".

Yow *et al.* (2000) identified three main reasons that organizations struggle to plan and it is the responsibility of the planning group to lessen these obstacles and encourage planning. First, lack of training and do not know how to plan. This lack of knowledge prevents them from planning and from expecting others in the department to plan. Second, some organizations do not think planning is necessary and see the process of planning as additional work without significant reward. These administrators fail to see the benefits of business planning. Third, firms see problems with the implementation of plans. These organizations may know how to plan and may know well the benefits of planning but don't believe implementation can be effective.

Trailor and Wolford (2001) concluded that, "listing the important topics in a business plan creates only generalities. Details must support the arguments made about and for the topics. That is, effective business plans are the outcome not only of covering all the bases, but also of covering them well. They further suggested that entrepreneurs may have significant issues writing meaningful business plans because of the details of writing substantive arguments. These issues may become even more significant, if the product or service is based on novel technologies that may not necessarily be easily understood by the market.

Reagan and Ghobadian (2002) stated that although the theories and frameworks for strategic planning are well-documented in academic crisis, the practical evidence of their application is in short supply. Beaver and Prince (2004) argued that planning in small business is more instinctive driven by intuition, informal methods of strategy formulation and implementation.

Ghobadian (2002) suggested in his study that small and medium sized firms have a less structured approach to strategic planning deployment.

Guerrero *et al.* (2012) aimed at identifying the main success factors for new ventures, efforts directed at evaluating the real effect of the existence and quality of a business plan on a firm's survival chances have been limited. The results of the study showed that none of the three variables that evaluate business plan quality including economic, financial and organizational viability seems to have a determining influence on survival chances.

Grant (2003) stated that in the last quarter of twentieth century, macroeconomic disequilibrium, exchange rate volatility, the microelectronics revolution and the emergence of newly industrializing countries have created inability to predict demand, prices, exchange rates and interest rates. Hence, the challenge in question influences companies' ability to plan. If established companies are to flourish and survive in the increased volatility business environment, strategies that are flexible and creative as well as less concerned with specific actions and more with establishing clarity of direction are required. The primary contribution of planning is to establish a process for strategic thinking and organizational learning.

Sevier (2003) supported the use of a process but noted that what one could do with the plan is more important than the plan itself. He cautioned against focusing so much on the process and the development of the plan that actions are forgotten. He further concluded that the goal is not the creation of a strategic plan, rather, the goal is a sense of direction and institutional coordination created by an effective strategic planning process. In other words, the plan is a guide to action".

Bryson (2004) reminded readers that business planning is "simply a set of concepts, procedures, and tools. Leaders, managers, and planners need to engage in strategic planning carefully because their success will depend at least in part on how they tailor the process to their situations".

Trainer (2004) indicated that much of the criticism of business planning is that "planning has focussed too heavily on the process of developing a plan rather than on implementation or outcome of the plan".

Gruber (2007) suggested that as many studies agree on was that business planning is heterogeneous task. Thus future research should take better heterogeneity into consideration that not all functional areas present in a business plan are planned in the same manner and they may not be of equal importance to venture success.

Lange *et al.* (2007) in their study found that firms with business plans had no performance differences from firms without business plans. They suggested that unless an entrepreneur needs to raise substantial start-up funds from venture capital or business angels, there is no compelling reason to write a detailed business plan before opening a new business.

Abdullah *et al.* (2008) commented on limitation faced by industries especially in the area of marketing of their services. Their study suggested on possible approaches for industries in exploiting marketing tools and techniques for their business success. According to them, government should play a vital role in educating the small scale industries. Also, the firms must be very creative and innovative in fully exploiting other marketing devices which are still acceptable by law to promote their services.

Gartner & Liao (2009) although pre-startup planning has been found to benefit new small businesses in several ways, there are still differing views on whether the process of pre-venture planning truly influences the survival of new small businesses. Many researchers argue the value of planning in terms of the time spent on planning, the usefulness of the information around the uncertain conditions, learning-survival linkage, and the relationship between planning and financial performance. The disadvantages of pre-startup planning of this review of the literature concentrate on the challenges that founders may confront, when planning their businesses. The most discussed disadvantage surrounds the possibility to plan in an unpredictable business environment.

Karlsson and Honig (2009) collected longitudinal data on six firms over a five year period to examine the impact of business planning on firm success. They found that firms initially conformed to business plan norms but over time they moved farther away from the plans. None of the Entrepreneurs who wrote business plans updated them and it was quite rare for an entrepreneur to refer to the business plans after they launched.

Brinckmann *et al.* (2010) hypothesized that in the early stages of the nascent entrepreneurial process, some planning oriented entrepreneurs may decide to abandon their venture earlier rather than later despite the legitimating effect discussed previously. While the theory and empirical evidence regarding planning and performance is somewhat inconsistent, there seems to be diminishing returns on business planning over time.

Ries (2011) planning fastened a decision-making process rather than trial-and-error learning can be also viewed from another angle. The claim stated that by planning founders can expand their knowledge and as follows to make better assumptions. Based on better assumptions, entrepreneurs can make mistakes on the paper rather than in action. However, there is a great possibility that the assumptions written in the plan turn out to be utterly flawed, if entrepreneurs have focussed solely on forecasting rather than gathering real feedback from the business environment. Hence, companies that focus on executing complex business plans may fail to notice vital issues that would have otherwise arisen from acting in the business environment. When companies in question have committed themselves so completely to act according to the assumptions made, they might not be able to adapt to the change in time after realizing the failure.

Features of business plan

Dhar and Lydall (1961) made their study on the data collected from Census of Indian Manufactures, 1956 and the study prepared by the Perspective Planning Division of the Planning Commission in respect of capital, labour and output relations in various industries. They concluded that the issue of choice between large and small industries for the purpose of an employment-oriented industrialization strategy is largely irrelevant, and it should aim at making the best use of scarce resources, instead of aiming at creating employment for the sake of employment.

McKelvie (1986) discussed business planning in terms of goal setting and the need for institutions to set clear goals in the face of economic and financial constraints. McKelvie stated that "Strategy involves outlining the institution's goals, their plans for achieving these goals, and the deployment of resources to attain these goals". She identified that strategic planning must have, as a foundation, a sound knowledge of the organization's mission. McKelvie also identified the results of business planning grounded in the organization's mission as: clarification and determination of the organization's long-range goals, selection of effective courses of action, and deployment of the appropriate allocation of resources necessary for the achievement of these goals.

Ansoff (1991) reasoned that planning generally produces better alignment and financial results than does trial-and-error learning. The intuitive appeal of these and similar arguments has led to the proliferation of academic and practitioner literature stressing the importance of planning, promoting models of the planning process, and offering normative advice on how to effectively design and implement strategic and operational plans.

Streib (1992) acknowledged that it is difficult to define the components of a successful strategic effort, but he identified four management functions that he deemed critical to the success of any business planning effort: leadership, human resources, managerial skills, and external support. The importance of people in the business planning process is evident in the fact that three of the four critical functions specifically address people and their role in planning.

Ruocco and Proctor (1994) suggested that an environmental analysis is a critical step in the planning process that must be performed to gather all the information necessary to develop appropriate strategies. The authors supported the use of SWOT (strengths, weaknesses, opportunities, threats) analysis as an effective and systematic way of matching the organization's strengths and weaknesses with the opportunities and threats that exist in the environment.

McGrath and MacMilan (1995) suggested that planning for new ventures is entirely different from planning for firms in other stages of development. These scholars argued that new ventures begin with a high ratio of assumption to knowledge and inevitably experience deviations from original targets that require fundamental re-direction. Thus, new ventures must practice more "discovery driven planning."

Wright (1995) supported the need in the business plan for efficiency by writing, "Change must address the need to become more efficient, and it should shift the focus of the organization from internal to external focus, to the customers and competing markets".

Castrogiovanni (1996) suggested that environmental uncertainty and founder knowledge should be further examined. In addition, future research could examine the attributes of good plans; which methods and procedures result in crafting a great plan.

Vinzant and Vinzant (1996) discussed opportunities and threats and present the notion that the analysis of opportunities and threats is not only an important part of the planning process but also the existence of these external forces is a catalyst for planning. Organizations are more likely to initiate and maintain strategic management systems because of the motivating factor presented by the existence of these opportunities and threats in the environment. The authors suggested that threats are in fact "often the only motivator powerful enough to generate sustained change on a large scale in complex organizations".

Poister and Streib (1999) characterized long-range planning as a "closed-system" orientation and contrasted this with the "big picture" approach of strategic planning that "emphasizes the importance of external trends and forces as they are likely to affect the agency and its mission"

Gilmore *et al.* (2001) reported networking as an inherent tool for business which is wholly compatible with SME decision-making characteristics in relation to marketing activities. It was illustrated that in case of competitor collaboration, it is balanced with caution when it is felt that the benefits enjoyed by each party are proportionate to their respective inputs and managers will guard against speaking too freely about certain things, especially any plans they might have for substantial changes to the firm or its direction.

According to Thompson and Strickland (2001) mission and vision are crucial elements of a business plan. The difference between vision and mission is in its time component; mission determines what the company does, whereas vision shows what the company will become and do in the future.

Sevier (2003) stressed that goals are the foundation for an effective plan that moves beyond paper to action, "A successful business plan, a plan that guides action is built on clear goals that are themselves built on solid data". This move to action, to work, is important or the plan will not be effective.

Delmar and Shane (2004) business planning can positively impact performance of new venture managers by influencing their motivation based on goal setting theory. However, goal setting theory functions best with respect to specific and simple (for the individual) goals. When goals are perceived as difficult and unspecific, as in the case of planning how to best run a business for the next five years, formal goal setting becomes detrimental. Because the goal setting involved in business planning is difficult (for the individual) and unspecific (determining the best strategy for developing a business opportunity over the next five years), entrepreneurs may, in fact, be rational to avoid making them.

Harding and Pines (2005) took up some of the problems that companies face. They do not give a broader perspective on the subject; they only scratch the surface a little bit. According to the authors, the solutions to the problems with the different types of risk are increased diversification and increased reserves. Companies should also focus more on the risks, but it is quite easy to say that they need to focus more and diversify more, but the authors give no examples of how this could be done.

Nichter and Golmark (2009) investigated factors associated with business growth i.e. (1) individual entrepreneur characteristics; (2) firm characteristics; (3) relational factors and (4) contextual factors. They further suggested that other researchers could keep in mind firms' opportunities and capabilities, as well as to other mechanisms by which factors potentially affect business growth.

Liao *et al.* (2009) suggested for a longitudinal research design, where individual firms are followed from birth to death, and then identified the individual, organizational, and context factors that may contribute to a firm's failure. "Researchers should make ongoing observations of the entrepreneurs, the environment, and the firm." Only through the longitudinal study with a quasi-experimental design approach would enable uncovering the factors that differentiate failed group from others. In addition, the moderating effects of a risk taking propensity should be examined in order to find out more about venture's competitive strategies.

Hakimpoor *et al.* (2011) hypothesized that the size, formality, diversity, density, stability and flexibility of marketing network will positively moderate the relationship between SMP and SMEs' performance to test the validity of model. The authors were intended to fill up the gap about the lack of research on SMP Performance relationship in SMEs' context. And they proposed that structural dimensions of marketing network are important moderators on relationship between strategic marketing planning (SMP) and SMEs' performance.

Chapter 3

Research Methodology

The research was conducted with the help of pre-tested schedule that incorporated information related to the perception of the entrepreneurs towards business plan, its challenges and the effective features on the success and the failure of a business. The information has been collected from the owners running small scale agro-based units in R.S.Pura and Bari-Brahmna areas of Jammu and Samba districts. The information collected has helped clearly in identifying the role of business plan.

Data collection

The data collection is the method to collect important information to keep on record for further use, to make important decisions about different issues, and will be of vital significance for others. The present study has adopted both the primary as well as secondary data collection techniques to conduct the present study:

- **Primary data**: The Primary data from the owners of the small scale agro-based units has been collected by survey method, using pre-tested structured schedule. Collection of data has also been done by the personal interview method of the owners so as to determine the importance of the business plan of small scale agro based units.
- **Secondary data**: The secondary data has been collected websites, journals, newspaper articles, annual reports of small scale units, magazines, etc.

Sampling technique

The present study has adopted two stage sampling technique, in which the first step was purposive sampling to conduct the survey among the different small scale agro-based units. Districts, blocks and then two circles were selected. Block R.S.Pura from Jammu district and Purmandal from Samba district were selected purposively as the maximum number of small scale industries are in these two areas. Then from these two blocks, R.S.Pura and Bari-Brahmna circles were selected purposively on the basis of maximum number of small agro-based units. The Second stage of sampling was simple random sampling for the selection of agro-based business units that was 25 each in Bari-Brahmna and R.S. Pura.

List of selected districts, blocks and circles

Two districts from Jammu and Kashmir state were selected for this study:

S.No.	Name of the districts	Name of the block	Name of the circle
1.	Jammu	R.S. Pura	R.S. Pura
2.	Samba	Purmandal	Bari-Brahmna

Circle-wise sample distribution

The sample distribution is as follows:

Circle	Number of units
R.S.Pura	25
Bari-Brahmna	25
Total number of units	**50**

Variables involved in the schedule

For objective 1: To find the perception of the entrepreneurs towards business plan, the data has been collected by asking questions from the owners of the units. The owners were supposed to answer the questions based on their locus of control, change in attitude, become innovative, etc. in their respective units the relevant data then collected were coded with "yes" (1) or "No" (2).

For objective 2: To analyze the challenges while running a business plan the owners were supposed to answer the questions based on various challenges like financial challenges, production challenges, time constraint, effective and technical labour, market challenges, high initial investment challenges, etc. The answers by the owners were given using likert-type scaling that they had to tick on any one of these: (strongly agree (1), agree (2), neutral (3), disagree (4) and strongly disagree(5) and the results are identified by using logistic regression.

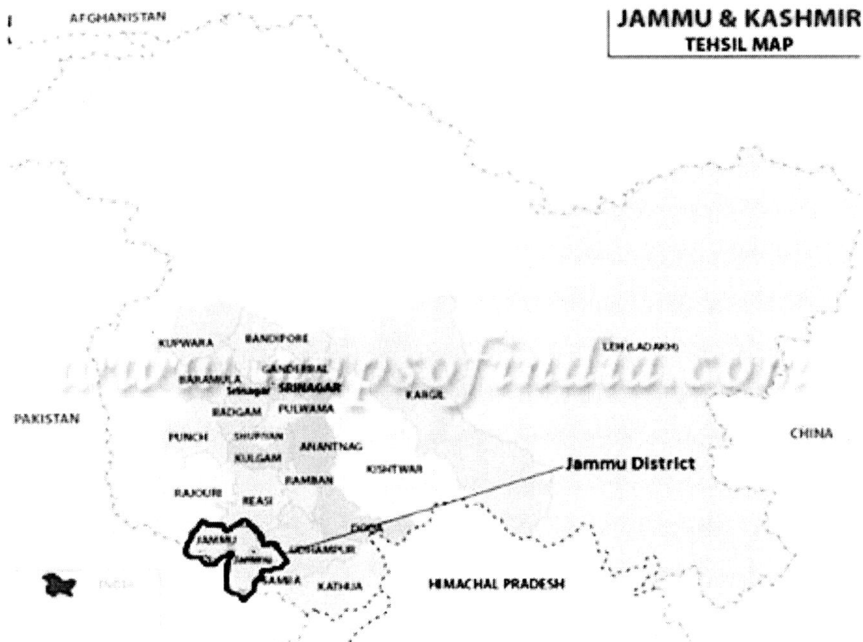

Fig. 1: Map of sample area Jammu

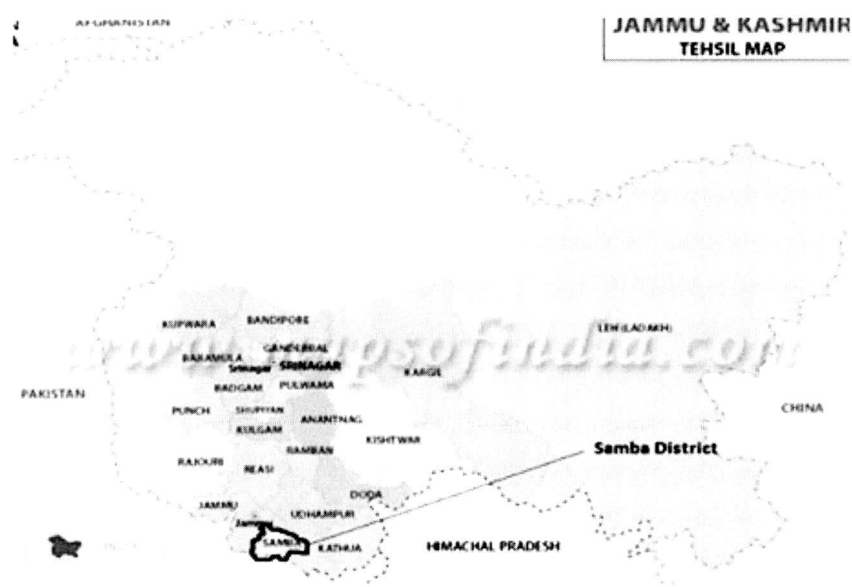

Fig. 2: Map of sample area Samba

For objective 3: Following variables have been taken into consideration to analyze the effect of essential features of a business plan. The owners had to rank the features of business plan from 1-12, where "1" denoted the most important and "12" least important. The features included were like clear and concise, well-organized, complexity of business, source of finance, technology upgradation, management, etc. and the results were identified using Garret ranking technique.

Methods of analyzing data

After collecting the data the results have been analyzed by using percentage analyses and descriptive statistics.

Percentage analysis

Percentage refers to a special kind of ratio. It is used to make comparison between two or more series of data. They can be used to compare the relative items, the distribution of two or more series of data, since the percentage reduces everything to a common base and there by allow meaningful comparisons to be made. In the present study the percentage analyses helped to check the effectiveness and the perception of entrepreneurs towards business plan.

Percentage = (x/y) X 100

Where, x = number of respondents respond

y = total number of respondents

Descriptive statistics

Descriptive statistics incorporate mean, median, standard deviation, that is used to analyze the variation between the factors and also helped to analyze the role of business plan in small scale agro-based units in Jammu.

Mean

Mean or average, in theory, is the sum of all the elements of a set divided by the number of elements in the set. Mean could be treated as a collaborative property of the whole set of values. Thus, the formula for mean will become.

Mean = Sum of all the set elements/Number of elements

The importance of mean lies in its ability to summarize the whole dataset with a single value.

$$mean = \overline{x} = (\Sigma Xi) / n$$

Where x_1, x_2 ....x_n, are the set of elements and n is the number of elements.

Logistic regression

Logistic regression measures the relationship between the categorical dependent variable and one or more independent variables, which are usually (but not necessarily) continuous, by estimating probabilities. Thus, it treats the same set of problems as does prohibit regression using similar techniques; the first assumes a logistic function and the second a standard normal distribution function.

Henry Garret ranking technique

In this technique, the percentage position of each rank, thus, obtained is converted into scores by referring to the table given by Henry Garret. Then for each factor the scores of individual respondents are added together and divided by the total number of respondents for whom the scores are added.

Formula

Percentage position= $100(R_{ij}-0.5)/n$

Where R_{ij} is the rank

n=number of items.

Results

The results pertaining to the present investigation **"Business Plan of Small Scale Agro-Based Units in Jammu"** have been presented in this chapter through appropriate tables and figures:

Out of the total number of respondents, 66 per cent answered that their enterprise have an proper business plan, but 34 per cent of the respondents answered that they do not have any business plan / or they do not need any business plan to start-up their business as is presented in Fig. 3.1.

Financial plan

Out of the total number of respondents, 62 per cent answered that their enterprise have a proper financial plan, but 38 per cent of the respondents answered that they did not have any financial plan / or they do not need any financial plan to start-up their business as is clear from Fig. 3.2.

Exit strategy

Figure 3.3, provides an insight of the total number of respondents from which 47 per cent answered that their enterprise have a proper exit plan and 53 per cent of the respondents answered that they do not have any exit plan/or they do not need any exit plan for their business.

Marketing plan

Figure 3.4 makes it clear that out of the total number of respondents, 46 per cent answered that their enterprise have a proper marketing plan. But 54 per cent of the respondents answered that they did not have any marking plan / or they do not need any marketing plan to start-up their business.

Does business plan play a critical role?

Figure 3.5, shows that out of total number of respondents, 64 per cent answered that business plan play a very critical role for the success of an enterprise, whereas 36 per cent of the respondents answered that business plan do not play a critical role for the success of an enterprise.

Are you having a written business plan?

Figure 3.6 shows that 30 per cent of respondents answered that they have written business plan, whereas 70 per cent of the respondents answered that they do not have written business plan.

Locus of control

In a stress full condition I am always at ease

The perusal of data from Fig. 3.7 makes it clear that 72 per cent feels easy while going through any stressfull situation as they can handle the problem in a very easy manner, remaining 28 per cent of the respondents admitted that they don't feel good in stressfull situation and feel very difficulty in making decision in stressfull situation.

I am able to manage stress properly

Figure 3.8 provides an insight that most of the respondents have showen that they can manage the stress in a very well manner, 74 per cent of the respondents have answered "yes".Only 26 per cent of the respondents felt difficult in managing the stress full situation and they also get stressed in that situation.

I am having control over my activities

Most of the respondents have shown that they have control over their activities, 78 per cent of the respondents have answered "yes".Only 22 per cent of the respondents felr that they didn't have control over their activities.

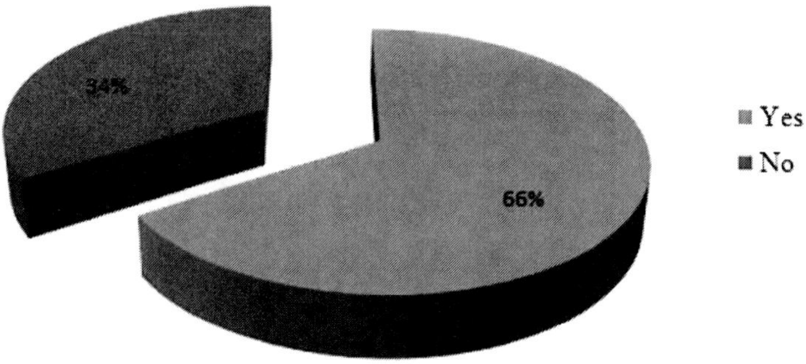

Fig. 3.1: Business plan

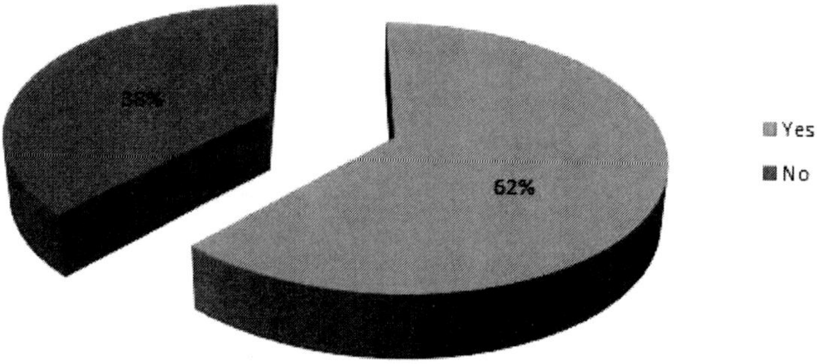

Fig. 3.2: Financial plan

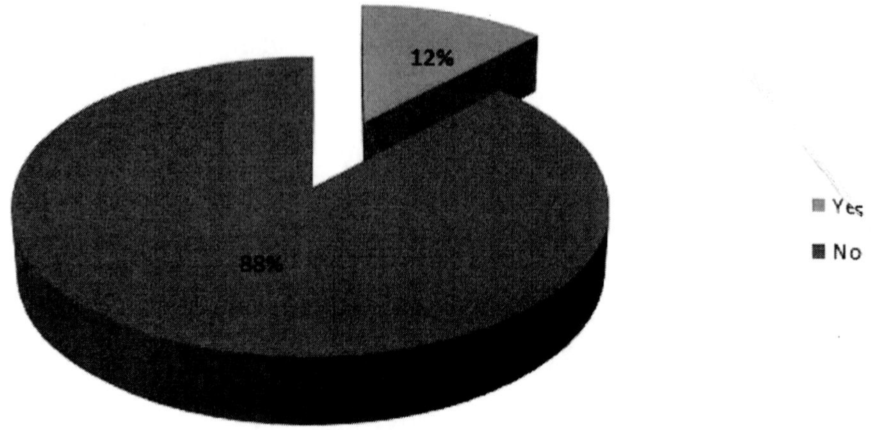

Fig. 3.3: Exit strategy

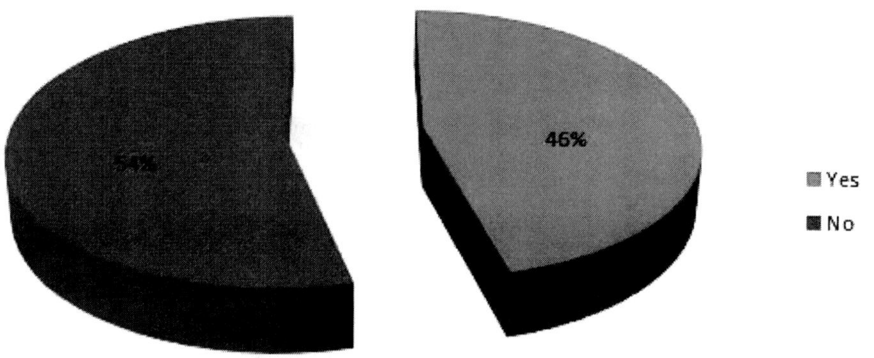

Fig. 3.4: Marketing plan

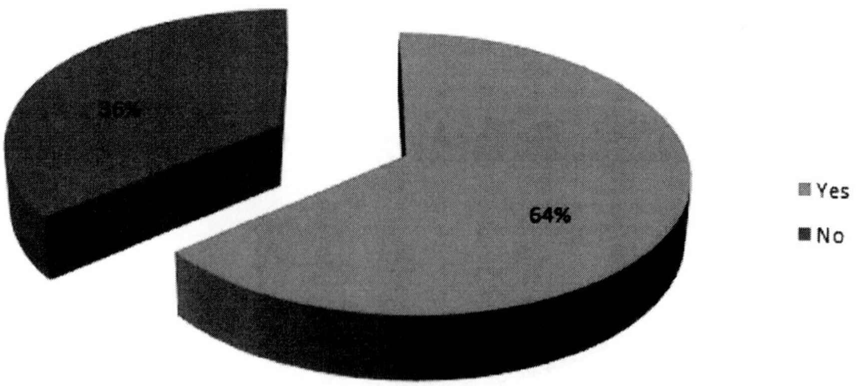

Fig. 3.5: Role of business plan

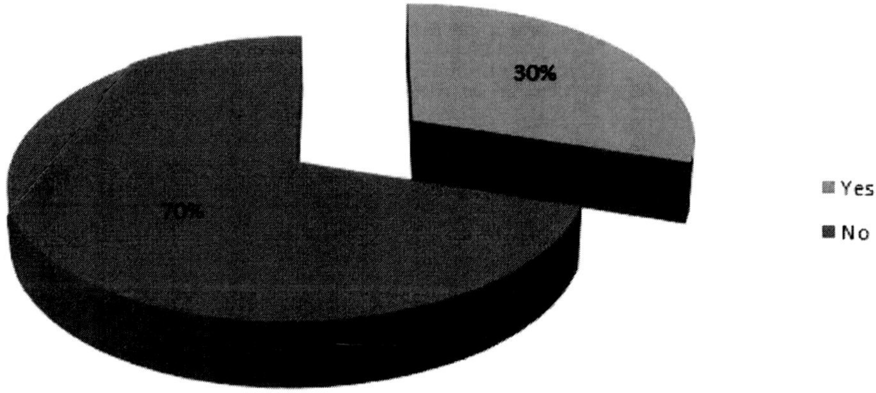

Fig. 3.6: Written business plan

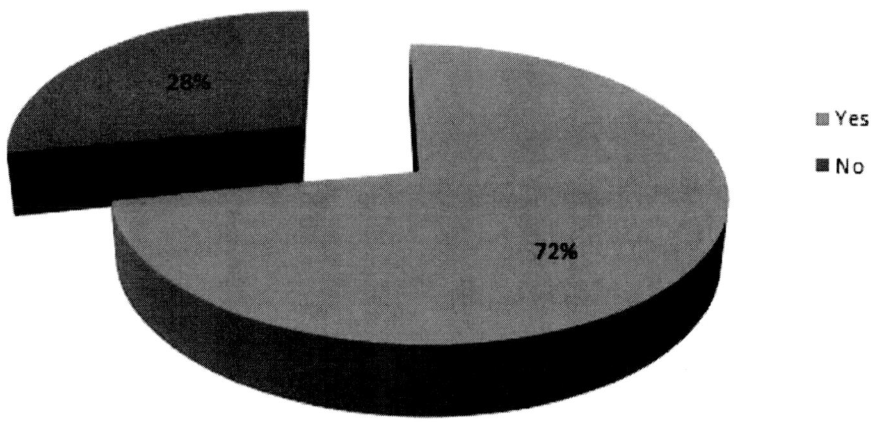

Fig. 3.7: Behaviour in stressful conditions

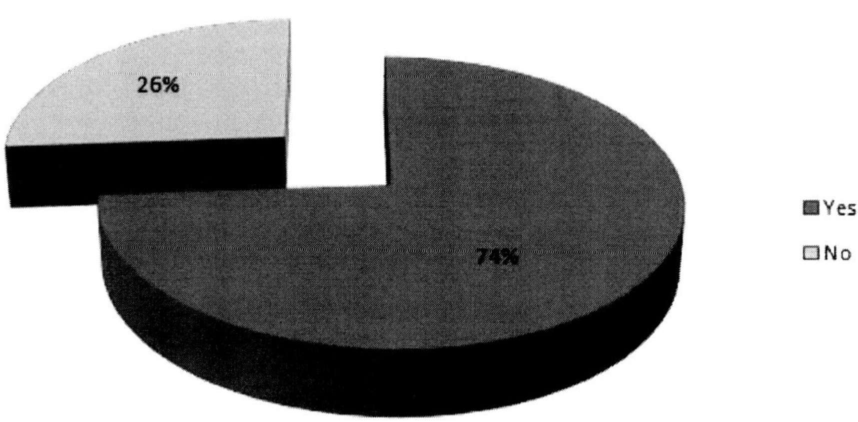

Fig. 3.8: Managing stress

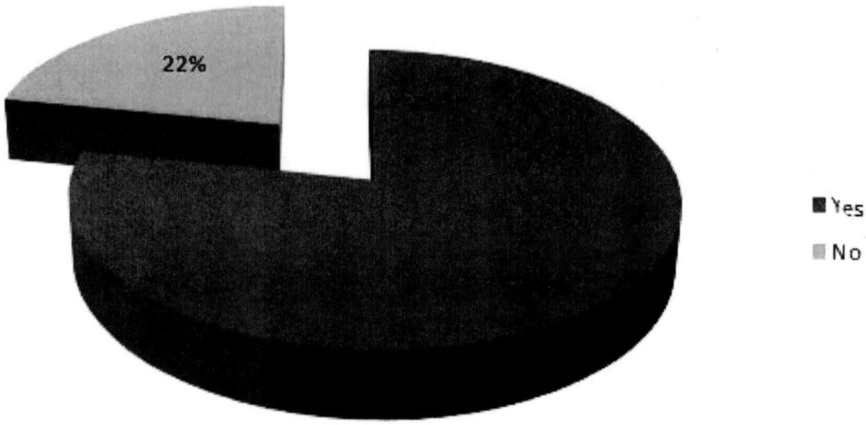

Fig. 3.9: Control over activities

Ambition

Figure 3.10, highlights 98 per cent of the total respondents were very ambitious in their work and only 2 per cent of the respondents admitted that they were not so much ambitious in their life. For a good business, the entrepreneur should have a quality of ambition.

Discipline

As per figure 3.11, 100 per cent of the respondents were very disciplined in their professional life._Respondents having discipline in their life can automatically have the ability to build a legal business plan in a disciplined manner.

Risk taker

Out of the total number of respondents, 52 per cent of the respondents were risk takers in their life, they take risky decision while performing in their business activities whereas remaining 48 per cent of the respondents admitted that they don't take risk in their business. Respondents having higher per cent will have the ability to build their business in the next level and play as a role model for others as is clear from fig. 3.12.

Positive attitude

Figure 3.13, highlights that 90 per cent of the respondents had positive attitude toward taking risky decisions or taking decisions in a very stress full situation. Remaining 10 per cent of the respondents had shown negative attitude in difficult situation and felt difficulty in making decision in stress full situation. Respondents having higher per cent of positive attitude will have the ability to build a good business plan.

Innovative

From the fig. 3.14, it is clear that 84 per cent of the respondents took innovative decision while performing their business activities .They believed in innovation rather than following traditional activities, 16 per cent of the respondents believed in following traditional activities rather than innovations and they felt uncomfortable in adopting new atmosphere. More innovative respondents can build more innovative business plan.

Self-efficacy

Figure 3.15 provides an insight that out of the total number of respondents, 88 per cent of the respondents were self-efficient means they did not relay on others for their activities, whereas 12 per cent of the respondents admitted that they were not self-efficient, they had to rely on others for their activities.

Was there a financial contribution provided to the enterprise by any financial institution?

100 per cent answered that they have taken financial help while starting up their enterprise as is clear from fig 3.16. As it is but natural that the finance cannot be made available unless and until the entrepreneur is able to show his business plan.

a) Bank

b) Other

Figure 3.17 shows 100 per cent of the respondents those answered yes have taken financial help from banks for starting-up their enterprise. So, the entrepreneurs knowingly or unknowingly had to prepare the business plan.

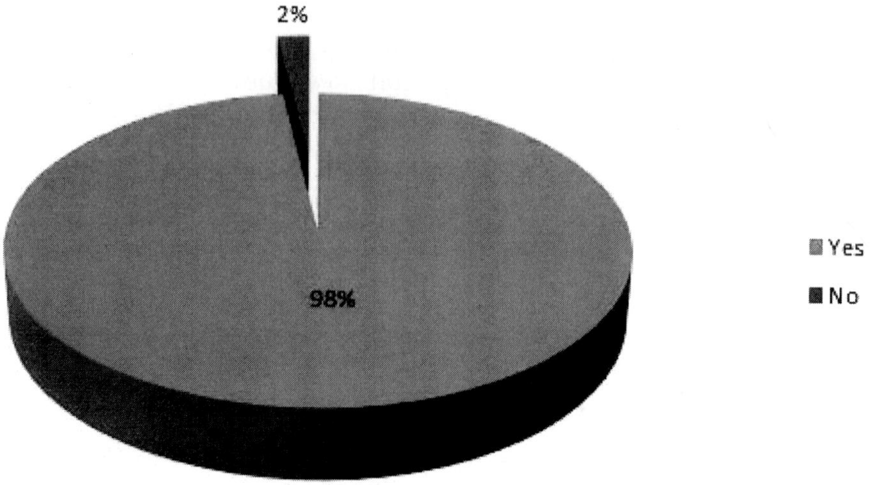

Fig. 3.10: Ambitious

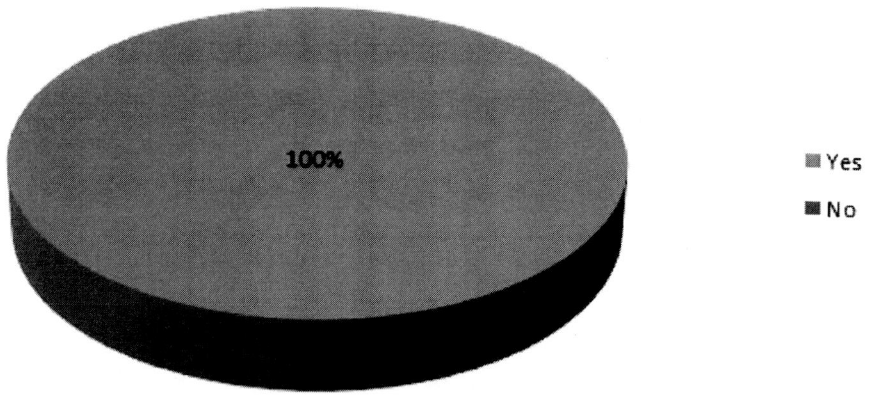

Fig. 3.11: Discipline

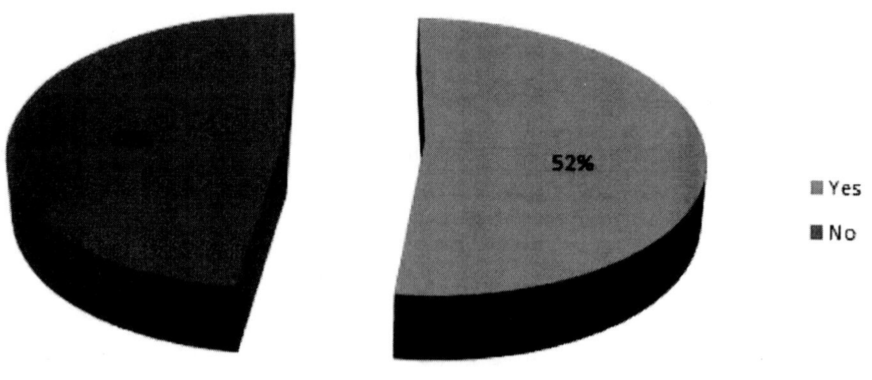

Fig. 3.12: Risk taker

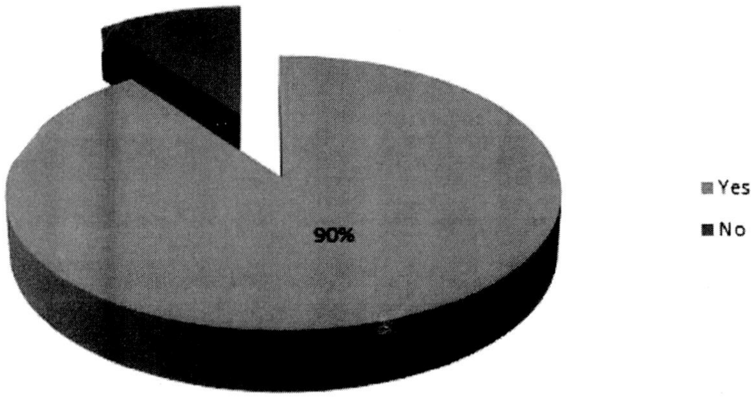

Fig. 3.13: Positive attitude

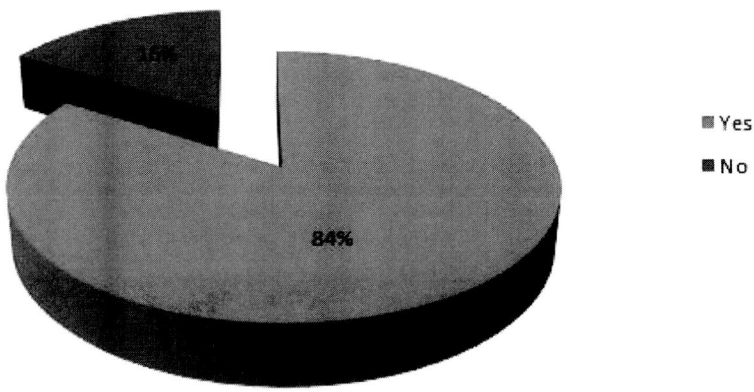

Fig. 3.14: Innovative

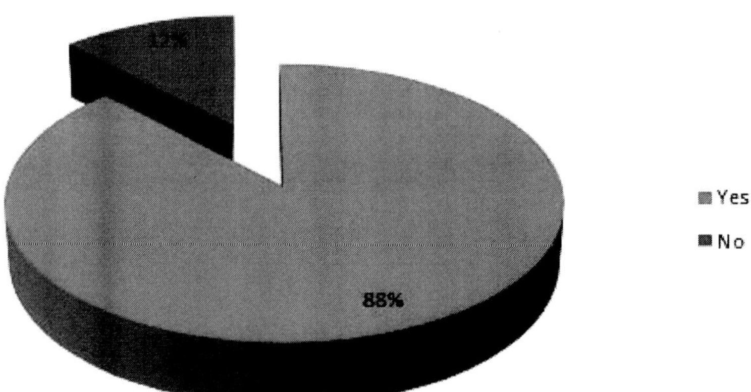

Fig. 3.15: Self-efficacy

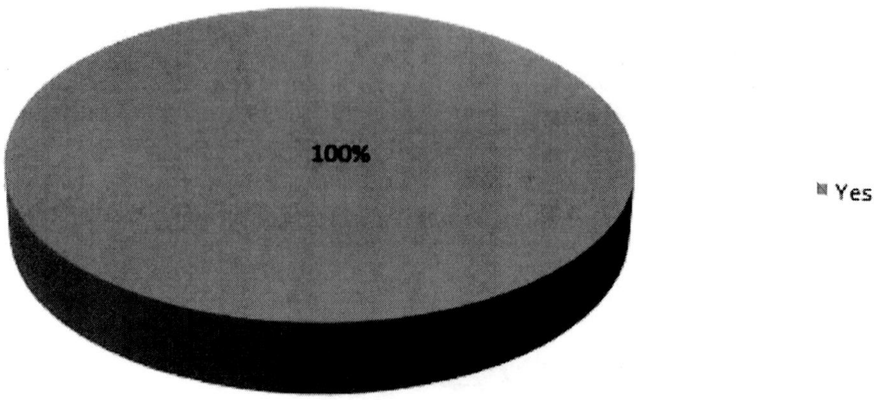

Fig. 3.16: Financial contribution

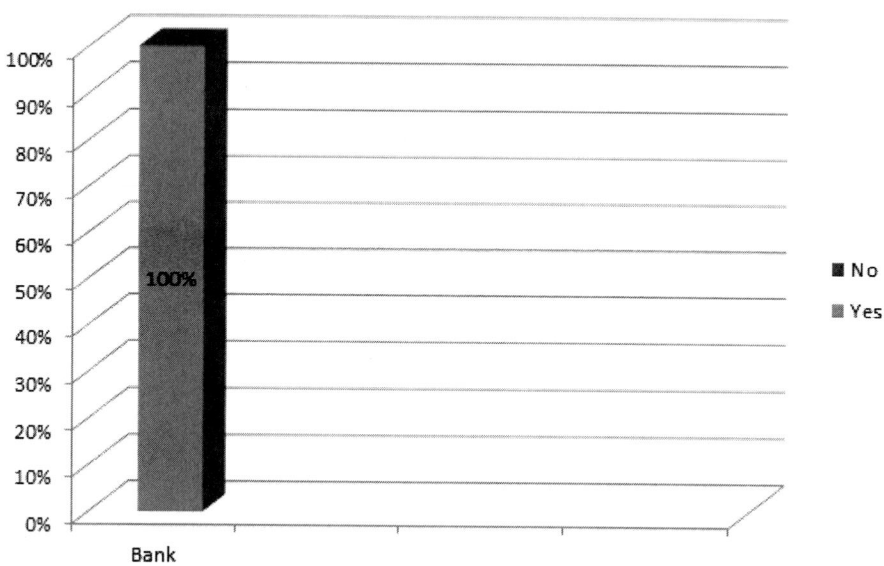

Fig. 3.17: Bank or other

During the issuing of loan provided by financing institute, they ask for business plan ?

Figure 3.18 shows that 44 per cent answered that yes the financial institute ask for a proper business plan and 56 per cent of the respondents answered that no the financial institute do not ask for any proper business plan from their enterprise.

Do they ask for the hard copy of business plan ?

34 per cent answered that yes the financial institute ask for a hard copy of business plan from these enterprises but 66 per cent of the respondents answered that no the financial institute do not ask for any hard copy of business plan from these enterprises as is clear from fig. 3.19.

Is there a proper re-evaluation of business plan since the opening of enterprise?

Out of the total number of respondents, 48 per cent answered that yes there is a proper re-evaluation of the business plan since opening of the business, whereas 52 per cent of the respondents answered that no there is no need of re-evaluation of the business plan.

What is probability time period that the organization undertake for Re-evaluation?

a) Monthly

b) Quarterly

c) Yearly

Figure 3.21 highlights only 2 per cent of the respondents go for re-evaluation of the business plan monthly, 26 per cent of the respondents showed that they go for re-evaluation of the business plan quarterly, whereas remaining 72 per cent of the respondents go for re-evaluation of the business plan yearly.

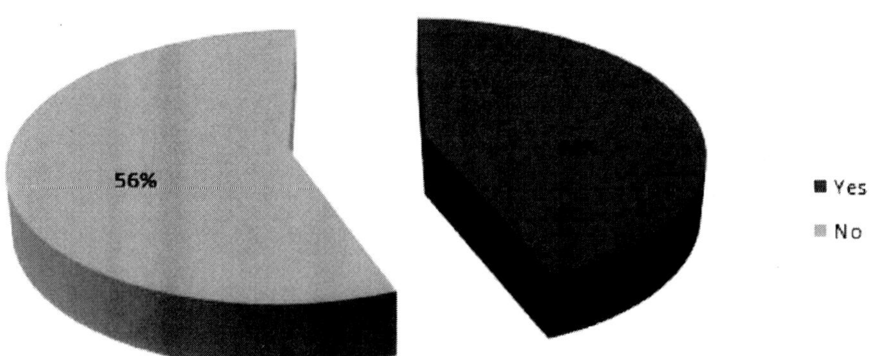

Fig. 3.18: Banks asking for business plan

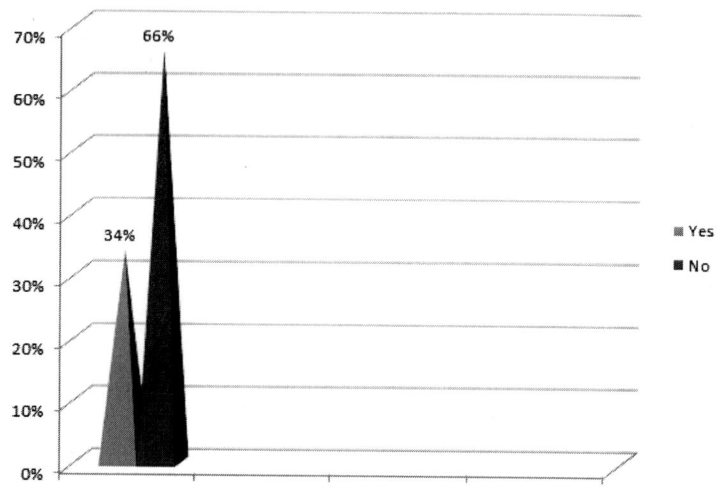

Fig. 3.19: Hard copy of business plan

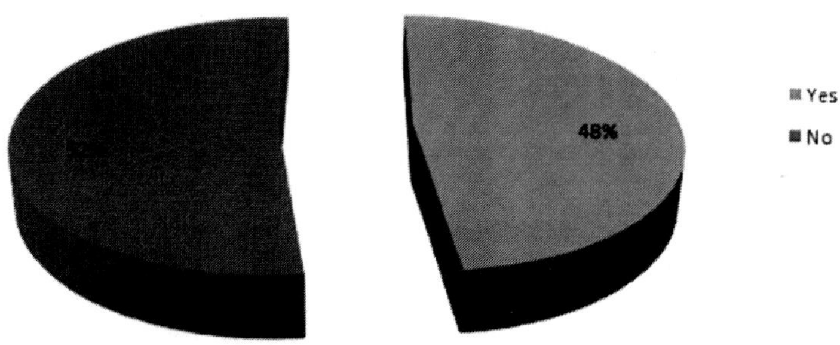

Fig. 3.20: Re-evaluation of business plan

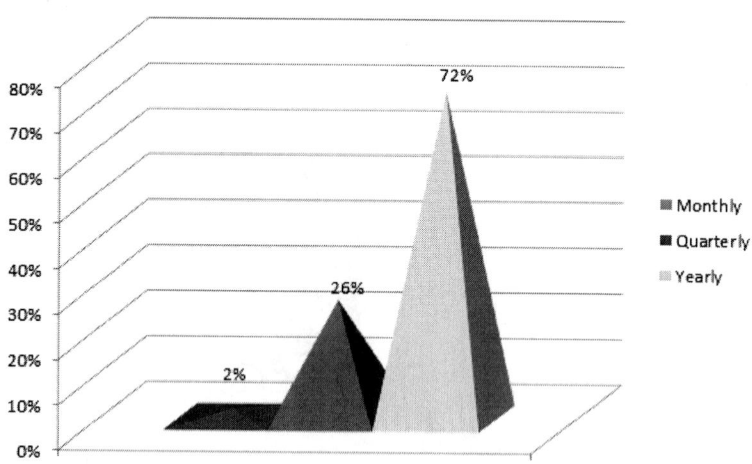

Fig. 3.21: Re-evaluation monthly, quarterly or yearly

Challenges while running a business plan

The success and the failure of an enterprise is critically dependent upon the challenges faced while running business activities. Table 3.3, highlights the role of different challenges in the success and failure of a business and concludes that for additional change in financial challenges the odds of success will be lower by a factor of 0.206. In other words, for additional improvement in financial challenges, the odd of surviving is lower by 79.4 per cent. For additional change in advertising challenges the odds of success will be lower by a factor of 0.012. In other words, for additional improvement in advertising challenges the odd of surviving is lower by 98.8 per cent. For additional change in other marketing challenges, the odds of success will be lower by a factor of 0.055. In other words, for additional improvement in other marketing challenges the odd of surviving is lower by 94.5 per cent. For additional change in production, time constraints and technical labour challenges the odds of success will be more because these challenges helps the firm to achieve their goals. For additional change in effective labour challenges, the odds of success will be lower by a factor of 0.195. In other words, for additional improvement in other effective labour challenges, the odd of surviving is lower by 80.5 per cent. For additional change in high initial investment challenges, the odds of success will be lower by a factor of 0.125. In other words, for additional improvement in high initial investment challenges the odd of surviving is lower by 87.5 per cent. For additional change in lack of government support challenges, the odds of success will be lower by a factor of 0.217. In other words, for additional improvement in high initial investment challenges, the odd of surviving is lower by 78.3 per cent. For additional change in time constraint challenges, the odds of success will be higher by a factor of 1.126. In other words, for additional improvement in time constraint challenges, the odd of surviving is higher by 83 per cent. For additional change in technical labour challenges, the odds of success will be higher by a factor of 8.660. In other words, for additional improvement in technical labour challenges the odd of surviving is higher by 13 per cent.

Table 3.1: Betterment of the business plan model

Model Summary			
Step	**-2 Log likelihood**	**Cox & Snell R Square**	**Nagelkerke R Square**
1	29.958[a]	.507	.695
a) Estimation terminated at iteration number 7 because parameter estimates changed by less than .001.			

Table 3.2: For determining the goodness of fit for logistic regression

Hosmer and Lemeshow Test			
Step	Chi-square	df	Sig.
1	4.313	8	.828

Table 3.3: Estimated regression coefficients of various factors and their standard errors

Variables	B	S.E.	Wald	df	Sig.	Exp(B)
A	-1.581	.810	3.807	1	.051	0.206
B	-4.462	1.664	7.187	1	.007	0.012
C	-2.899	1.119	6.716	1	.010	0.055
D	.532	.632	.709	1	.400	1.702
E	.315	1.145	4.089	1	.043	1.126
F	-1.633	.896	3.319	1	.068	0.195
G	2.159	1.004	4.620	1	.032	8.660
H	-2.078	.911	5.204	1	.023	0.125
I	-1.529	.899	2.893	1	.089	0.217
Constant	18.574	6.982	7.077	1	.008	1.166

Note: A = Financial challenge; B = Advertising challenge; C = Other marketing challenge; D = Production challenge; E = Time constraint; F = Effective labour; G = Technical labour challenge; H = High initial investment challenge; I = Lack of government support

Essential features for an effective business plan

Table 3.6, provides an insight about the ranking of the essential features of an effective business plan. Thus, from the total number of respondents the average rank given to the essential features are that Clear/ concise has occupied 1st rank, well-organized 2nd rank, source of finance are clear has occupied 3rd rank, proprietary position 4th rank, technology upgradation 5th rank, market orientation has occupied 6th rank, preference of the management team 7th rank, procurement procedure and policy 8th rank, level of uncertainty has occupied 9th rank, management style and ability of the management has occupied 10th rank, proper upgrading of human resource 11th rank, complexity of a business has occupied 12th rank.

Table 3.4: Average response of respondents for ranking of different features

	1	2	3	4	5	6	7	8	9	10	11	12
A	0	0	0	2	2	7	3	10	3	8	10	5
B	4	0	2	1	5	1	8	3	8	5	5	8
C	0	0	0	0	1	2	5	8	13	6	10	5
D	0	0	0	0	5	5	5	11	6	10	2	6
E	12	17	12	7	1	1	0	0	0	0	0	0
F	12	13	11	5	7	0	2	0	0	0	0	0
G	1	1	0	0	7	9	7	8	5	1	9	2
H	7	7	7	13	7	4	1	1	0	1	1	1
I	11	7	10	4	7	2	2	5	2	0	0	0
J	0	0	1	7	2	7	6	6	1	8	2	10
K	2	5	7	6	4	6	4	0	5	7	4	0
L	1	0	0	4	2	7	6	4	4	1	7	14

Table 3.5: Garret score table

Rank	1	2	3	4	5	6	7	8	9	10	11	12
Score	4.2	12.5	20.8	29.2	37.5	45.8	54.2	62.5	70.8	79.2	87.5	95.8
Garret Score	83	73	66	60	56	52	48	44	39	34	27	17

Table 3.6: Ranking of essential features of an effective business plan

	1	**2**	**3**	**4**	**5**	**6**	**7**	**8**	**9**	**10**	**11**	**12**	**TS**	**AS**	**Rank**
A	0	0	0	120	112	364	144	440	117	272	270	85	1924	38.48	X
B	332	0	132	60	280	52	384	132	312	170	135	136	2125	42.5	VII
C	0	0	0	0	56	104	240	352	507	204	270	85	1818	36.36	XII
D	0	0	0	0	280	260	240	484	234	340	54	102	1994	39.88	IX
E	996	1241		792	420	56	52	0	0	0	0	0	0	3557	71.14
I	913	511	660	240	392	104	96	220	78	0	0	0	3214	64.28	III
F	996	949	726	300	392	0	96	0	0	0	0	0	3459	69.18	II
G	83	73	0	0	392	468	336	352	195	34	243	34	2210	44.2	VI
H	581	511	462	780	392	208	48	44	0	34	27	17	3104	62.08	IV
I	913	511	660	240	392	104	96	220	78	0	0	0	3214	64.28	III
J	0	0	66	420	112	364	288	264	39	272	54	170	2049	40.98	VIII
K	166	365	462	360	224	312	192	0	195	238	108	0	2622	52.44	V
L	83	0	0	240	112	364	288	176	156	34	189	238	1880	37.6	XI

❑❑❑

Conclusion

1. Perception of entrepreneurs towards the importance of business plan

Majority of the owners of the small scale agro-based units in Jammu and Samba districts of J&K state finds business plan helpful in managing stress properly, controlling activities, setting of an ambition, builds discipline, generates risk taking ability, generate positive attitude and with the help of business plan they find themselves self-efficate.

2. Challenges while running a business plan

The challenges like financial challenges, advertising challenges, other marketing challenges, effective labour challenges, high initial investment challenges, lack of government support challenges have negative relationship with the success of a business plan, whereas challenges like time constraint, production, technical labour have positive relationship with the success of a business plan.

3. Effective features of a business plan

The features like clear/concise, well-organized, source of finance, proprietary position, technology upgradation, market orientation are some most important or effective features of a business plan, whereas complexity of a business, proper upgrading of human resource, preference of the management team, procurement procedure and policy, level of uncertainty are the least important features of a business plan.

Conclusion

This study shows that 66 per cent of the small scale businesses have a proper business plan and 34 per cent have no business plan while running their business operations and activities. This shows that there is a need of guiding these small scale businesses so that they get the knowledge about the managerial skills, planning and how to take managerial decisions. The second proposition on the challenges faced by the small enterprises like financial challenges, advertising challenges and other marketing challenges are having negative impact on the success of an small scale businesses. They have to manage these challenges by taking loans from the banks/financial institutes and by gaining the knowledge of managerial skills. The data also supported the proposition that the features of a business plan have a positive impact on the success of a business. It also shows that market orientation, proprietary position, procurement procedure and policy are some of the features that have high impact on the success of a small agro-based business.

Policy Implications

Business planning remains a critical element of managing a business. The process of business planning allows the business owner to analyze the competitive environment of the business as well as its internal capabilities enabling one to position the business correctly. These are some recommendations given while analysing the data :-

- There is a need to educate the entrepreneurs regarding the benefits associated with the business plan for the success of their enterprises.
- Entrepreneurs have to do the research of the market while making a business plan. They have to think from all directions so that no area remains uncovered.
- Small entrepreneurs have to go for loans from banks for their initial investment and submit their business plan in the bank, the acceptance of business plan will show the future growth of their business.
- It has been found that there is a great impact of the features of the business plan on success or failure of business plan, hence it is recommended that while making a business plan, the entrepreneur has to give focus on the features and challenges of business plan.

❑❑❑

Bibliography

Anonymous, (2013). Annual Survey of Industries, Central Statistical Organization, New Delhi, India.

Anonymous, (2014). jandkplanning.com/economic-survey.

Armstrong, (1982). The Value of Formal Planning for Strategic Decisions: Review of Empirical Research. *Strategic Management Journal*, 3: 197–211.

Ansoff, I. (1991). "Critique of Henry Mintzberg's 'The Design School: Reconsidering the Basic Premises of Strategic Management." *Strategic Management Journal*, 12: 449–461.

Abdullah, A., Kamarulzaman, Y., and Farinda, A., G. (2008). 'Legal Services and Marketing Limitations: A focuson SME in Malysia, 4(5): 308-319.

Acs, Z.J., Arenius, P., Hay, M., & Minniti, M. (2005). Global entrepreneurship monitor: 2004 executive report. Babson Park, MA: Babson College, and London: London Business School.

Bracker, *et al.* (1988). Planning and financial performance among small firms in a growth industry. *Strategic Management Journal*, 9(6): 591-603.

Bruderl, J., & Schussler, R. (1990). Organizational mortality: *The liabilities of newness and adolescence.* Administrative Science Quarterly, 35(3): 530–547.

Bhide, A. (1994). How entrepreneurs craft strategies that work. *Harvard Business Review* 72(2): 150-161.

Bryson, J.M. (2004). Strategic planning for public and non-profit organizations: a guide to strengthening and sustaining organizational achievement (3rd ed.). San Francisco: Jossey-Bass.

Brinckmann, J., Grichnik, D. and Kapsa, D. (2010). Should Entrepreneurs Plan or Just Storm the Castle? A Meta-analysis on Contextual Factors Impacting the Business Planning–performance Relationship in Small Firms. Journal of Business Venturing, 25(1): 24–40.

Castrogiovanni, G. (1996). Pre-Startup planning and the survival of new small businesses: Theoretical linkages. *Journal of Management,* 22(6): 801-822.

Dhar, P.N. and Lydall, H. (1961). The Role of Small Enterprises in India's Economic Development, Asia Publishing House, Bombay.

Drucker, P.F. (2001). *Management challenges for the 21st century.* New York: Harper Business.

Delmar, F. & Scott, S. (2003). Does business plan facilitate the development of new ventures? *Strategic Management Journal,* 24(11): 1165-1185.

Delmar, F. and S. Shane. (2004). Legitimating first: Organizing activities and the survival of new ventures, *Journal of Business Venturing* 19(3): 385-410.

Floyd, S.W. & Wooldridge, B. (1996). Middle management's strategic influence and organizational performance. *Journal of Management Studies,* 34(3): 465-485.

Franco, M. & Haase, H. (2010). Failure factors in small and medium-sized enterprise Qualitative study from an attributional perspective. *International Entrepreneurship and Management Journal,* 6(4): 503–521.

Gilmore, A., Carson, D. and Grant, K. (2001), "SME marketing in practice", Marketing Intelligence & Planning, 19(1): 6-11.

Grant, R.M. (2003). Strategic planning in a turbulent environment: Evidence from the oil majors. *Strategic Management Journal,* 24(6): 491-517.

Gruber, M. (2007). Uncovering the Value of Planning in New Venture Creation: A Process and Contingency Perspective. *Journal of Business Venturing,* 22: 782–807.

Gartner, W.B. & Liao, J. (2009). Report for Small Business Association, Office of Advocacy (2009). Are planners doers ? Pre-venture planning and the start-up behaviors of entrepreneurs.

Harding, F. and Pines, G. (2005): *Lessons in revenue Risk Management,* Consulting to Management, 16(2): 39-42.

Hakimpoor, H., Tat, H.H. and Arshad, K.A. (2011). "Strategic Marketing Planning (SMP) and SMEs' Performance: The Moderating Effects of Structural Dimensions of Marketing Networks" 2nd International Conference on Business and Economic Research (2nd ICBER 2011) Proceeding, 1013-1025.

Khusro, A.M. (1999): The Poverty of Nations, Palgrave Macmillan, London.

Karlsson, T. and Honig, B. (2009). Judging a Business by its Cover: An Institutional Perspective on New Ventures and the Business Plan. *Journal of Business Venturing*, 24: 27–45.

Lane, R.J., Bishop, H.L. & Wilson-Jones, L. (2005). Creating an effective strategic plan for the school district. *Journal of Instructional Psychology*, 32(3): 197–204.

Liao, J. and Gartner, W. (2006). The Effects of Pre-venture Plan Timing and Perceived Environmental Uncertainty on the Persistence of Emerging Firms. Small Business Economics, 27(1): 23-40.

Liao, J., Moutray, C. and Welch, H. (2009). Start-up resources and entrepreneurial discontinuance: The case of nascent entrepreneurs. Journal of Small Business Strategy, 19(2): 1-15.

Lange, J.E., Mollov, A., Pearlmutter, M., Singh, S. and Bygrave, W.D. (2007). Pre-start-up Formal Business Plans and Post-start-up Performance: A Study of 116 New Ventures. Venture Capital: *An International Journal of Entrepreneurial Finance*, 9(4): 237-256.

Mintzberg, H. (1978). *Patterns in strategy formation*. Management Science, 24(9): 934-948.

McKelvie, B.D. (1986). The university's statement of goals: an idea whose time has arrived. Higher Education, 15(1/2): 151-163.

Mintzberg, H. (1990). The design school: Re-considering the basic premises of strategic management. Strategic Management Journal, 11(3): 171-195.

Mintzberg, H. (1994). The rise and fall of strategic planning. New York: The Free Press.

McGrath, R.G., and MacMillan, I.C. (1995). "Discovery-Driven Planning." *Harvard Business Review*, 73(4): 44–54.

Mc Carthy, B. (2003). Strategy is crisis-driven: insight from entrepreneurial firms. Management Decision, 41(4): 327-339.

Nichter, S. and Goldmark, L. (2009), "Small Firm Growth in Developing Countries", World Development, 37(9): 1453-1464.

O'Reagan and Ghobadian, (2002). Effective strategic planning in small and medium-sized firms. Management Decision, 40(7): 653-671.

Poister, T.H. & Streib, G.D. (1999). Strategic management in the public sector: concepts, models, and processes. Public Productivity Management Review, 22(3): 308-325.

Ruocco, P. & Proctor, T. (1994). Strategic planning in practice: a creative approach. Marketing Intelligence and Planning, 12(9) : 24-29.

Ries, E. (2011). The lean startup. (1st ed.) Crown Publishing Group, New York.

Streib, G. (1992). Applying strategic decision making in local government. Public Productivity & Management Review, 15(3): 341-354.

Sevier, R.A. (2003). From strategy to action. University Business, February, 18-19.

Shane, S. and Delmar, F. (2004). Planning for the Market: Business Planning before Marketing and the Continuation of Organizing Efforts. *Journal of Business Venturing*, 19: 767-785.

Shepherd, D., and J. Wiklund (2009). "Are We Comparing Apples with Apples or Apples with Oranges? Appropriateness of Knowledge Accumulation across Growth Studies," Entrepreneurship: Theory & Practice, 33(1): 105-123.

Thompson, A.A., and A.J. Strickland. (2001). *Strategic Management: Conceptsand cases.* (12th ed.) Boston, MA: McGraw-Hill/Irwin.

Trainer, J.F. (2004). Models and tools for strategic planning. New Directions for Institutional Research, 123: 129-138.

Trailor, J. and Wolford, C. (2001). On the Logic of Business Plan Composition. *New England Journal of Entrepreneurship*, 4(1): 41-50.

Vinzant, D.H., & Vinzant, J.C. (1996). Strategy and organizational capacity: finding a fit. Public Productivity & Management Review, 20(2): 139-157.

Wilkinson, G. & Monkhouse, E. (1994). Strategic planning in public sector organizations. Executive Development, 7(6): 16-19.

Wright, C.C. (1995). Strategic planning for higher education. *Journal of Professional Issues in Engineering Education and Practice*, July: 187-190.

Yow, D.A., Migliore, R.H., Bowden, W.H., Stevens, R.E. & Loudon, D.L. (2000). Strategic planning. New York: The Hawthorn Press, Inc.

□□□